100 BILLION SUNS

100 BILLION SUNS

The Birth, Life, and Death of the Stars

RUDOLF KIPPENHAHN

Translated by Jean Steinberg

Princeton University Press
Princeton, New Jersey

Published by Princeton University Press, 41 William Street,
Princeton, New Jersey 08540

The Princeton Science Library paperback edition
is published by arrangement with Basic Books,
a division of HarperCollins Publishers, Inc.

Kippenhahn, Rudolf, 1926–
[Hundert Milliarden Sonnen. English]
100 billion suns: the birth, life, and death of the stars /
Rudolf Kippenhahn; translated by Jean Steinberg.
p.   cm.—(The Princeton science library)
Originally published: New York: Basic Books, © 1983.
With new afterword.
Translation of: Hundert Milliarden Sonnen.
Includes index.
1. Stars—Evolution. I. Title. II. Title: One hundred billion suns. III. Series.
[QB806.K5313   1992]   92-15753   523.8′8—dc20

ISBN 0-691-08781-4 (pbk.)
First Princeton Paperback printing, 1993
1   3   5   7   9   10   8   6   4   2

Contents

CONTENTS

Preface to the
Paperback Edition

This book grew out of more than a hundred lectures given to lay audiences about the exciting world of astrophysics, and it took shape in the fall semester of 1978 in the course of a series of university lectures before a general student audience at the University of Munich. Portions of the text closely follow some of the essays Alfred Weigert and I wrote for the journal *Sterne und Weltraum*. Some of the material is interlaced with personal reminiscences, since much of what I am writing about has become known only during the past twenty-five years. I myself "experienced" these discoveries, and in some instances some of my friends and I had the good fortune to be personally involved in them.

In 1981 the news that this book was to appear in the United States filled me with deep joy. My thoughts went back to that day in 1961 when the ship that brought me to the United States for the first time docked in Hoboken, New Jersey.

That visit was to be a turning point in my life. In looking back on my professional career, I feel that it can be divided into two parts: pre–United States and post–United States. The work I did after my visit was freer, less constrained, stamped as it were by my association with Princeton and Pasadena. But professional enrichment was not the only profitable part of my stay. No less significant were the colleagues I met and the friends I made. I am thinking in particular of Robert E. Danielson (Princeton, New Jersey), Louis G. Henyey (Berkeley, California), and Marshall H. Wrubel (Bloomington, Indiana), valued friends and colleagues of that time who are no longer with us.

Since the American hardcover edition came out in 1982 it is necessary to add a few remarks:

The X-ray source Hercules X-1 (p. 175–180) has meanwhile shown peculiar behavior. Its X-ray radiation faded in May 1983, and was almost undetectable until March 1984. During this time the regular brightness variations of Hoffmeister's star were unchanged. This suggests that the X-ray source was active all the time and its radiation was striking Hoffmeister's star, but that for some reason we were shielded from the X-rays.

In 1983 Professor Chandrasekhar (p. 197, 198) received the Nobel prize in physics for his work on white dwarfs.

The discrepancy between the theoretical solar models and Prof. Davis's solar neutrino experiment (p. 88, 89) remains still unresolved. With respect to the Gallium experiment (p. 90, 91), the sensitivity of receivers has been improved so that only thirty tons of Gallium are needed. At the moment Italy, France, Israel, and the German Federal Republic are planning to carry out the experiment in a tunnel in the Gran Sasso Massive in Italy.

I want to thank the personnel of Basic Books for preparing the paperback edition. At the European end I had the help of my colleagues Walter F. Huebner, Ronald E. Kates, Michael L. Norman, and Martin A. Walker, who went through the translation and suggested several improvements.

It is my hope that readers in the land that has given me so much will share some of the pleasure I had in writing this book.

Rudolf Kippenhahn
Munich, November 1985

100 BILLION SUNS

Introduction

The scene of action is the Milky Way. The dramatis personae are the 100 billion stars of the Milky Way and a few hundred earthbound astronomers.

Following a script drawn up by the laws of nature, matter has gathered into bodies that we call *stars*. Neither solids nor liquids can survive their great heat. Stars are thus spheres of gas held together by the force of their own gravity. We refer to one such sphere as the sun. To an observer outside our galaxy the sun would be just another medium-sized star among 100 billion others in the Milky Way—neither particularly large nor particularly small, of average brightness, and without any special significance. But for us who derive our life from it, the sun is very important indeed.

Most of the stars that make up the Milky Way are situated in our *galaxy*, a round, shallow disk so vast that it takes light almost 100,000 years to traverse it diagonally from rim to rim. The disk rotates, and all the stars move around its center, forced into complicated patterns by the interplay of centrifugal and gravitational forces.

Ours is not the only star system in the cosmos. The Andromeda nebula is another rotating disk of stars (see plate I). Because this nebula is seen at an angle, the disk appears to be elliptical. The Andromeda nebula is a replica of our own system. In it are found the same types of stars and the same processes as in our system, and not only there, for there exist thousands, millions, perhaps even an infinite number of other galaxies.

In plate II we are given a vertical view of another star system. That the Milky Way and the remote, often spiralform nebular objects in the sky are of the same type was not established unequivocally

3

Figure I.1 The example of the Andromeda galaxy demonstrates how the ribbon of the Milky Way is formed. An observer looking from his planet within the disk outward beyond its plane would see something like the illustration on the upper left: comparatively few stars would cross his line of sight. A view along the plane of the disk would offer the picture at the upper right: a luminous ribbon stretching across the sky. (Photograph: U.S. Naval Observatory, Washington, D.C.)

until 1924. The presence of the small, faint, frequently elliptical nebular disklets called *spiral nebulae* had, however, been recognized for a long time. As far back as 1755, the then thirty-one-year-old Immanuel Kant, in his *Universal Natural History and Theories of the Heavens,* had already compared them with our own stellar system: "If such a world of fixed stars [Kant was referring to our Milky Way] were looked at from such an immeasurable distance by an observer standing outside it, subtending a small angle, it would appear as a poorly lit smudge of circular shape when its surface offers itself up to the eye directly, and elliptical if seen sideways." Kant concluded that the elliptical nebulae in the sky were distant Milky Way systems: "Everything speaks for taking these elliptical figures for just such

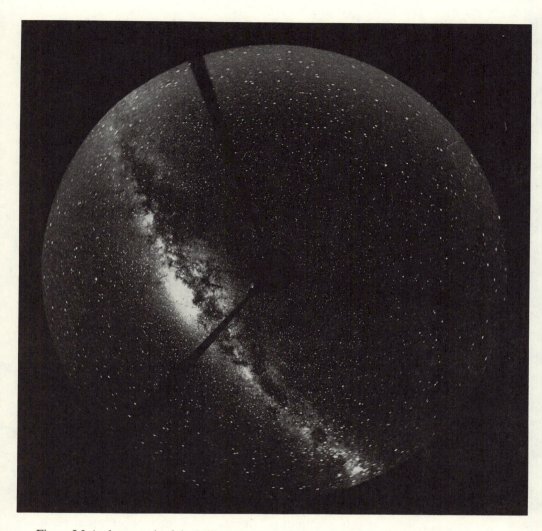

Figure I.2 A photograph of the Milky Way taken with a wide-angle lens. The dark wedges seen in the picture are caused by the camera. (Photograph: W. Schlosser, Astronomical Institute, Ruhr University, Bochum, German Federal Republic)

cosmic systems, for Milky Ways, whose constitution we have just described." Yet another two hundred years were to pass before this could in fact be proved.

The sun, and we along with it, are situated almost in the central plane of the Milky Way. When we look outward vertically from the

5

disk into space, we can see only a comparatively few stars, but when we look toward the rim, we can see a great many, as figure I.1 demonstrates. The flat disk of our stellar system shows up as a luminous ribbon stretching across the night sky—the ribbon of the Milky Way, as seen in figure I.2.

But not only stars crowd the disk of our stellar system. Luminous clouds float and fill in the spaces between the stars. As much as 1/100th part of the mass of the Milky Way is concentrated not in the stars of that system but in the spaces between the stars. The chemical composition of this mass is similar to that of the sun but is only a millionth of a billionth of a billionth as dense. Minute dust particles with radii of only 1/10,000th of a millimeter are embedded in this *interstellar* gas. Like heavy curtains interstellar dust clouds dilute the light of the stars behind them and give it a reddish hue, just as the atmospheric dust of the earth imparts a red tinge to the setting sun.

The stars, gas, and dust of the Milky Way move sluggishly; it takes them about 100 million years to complete their orbit around the center of the system. But though their orbit is slow the stars are not at all inactive. Many form binary systems in which two stars revolve around each other in a period of years, days, or even hours. Others expand and contract in a rhythmic pattern, as though breathing. From time to time a star will explode and for a brief period become as luminous as all the 100 billion stars of the Milky Way put together. Some stars do not radiate their light steadily but give off flashes of light one after the other in intervals of hundredths of seconds.

Looking at this vast natural drama from their observation posts on the minute planet Earth as it revolves around the insignificant star called the sun, a handful of astronomers seek to gain an understanding of the cosmos. Using instruments constructed from materials found on their planet, they follow the activities in space from their observatories and launch rocket-borne telescopes from Earth. Some people confuse them with astrologers, but astronomers reject all such notions of kinship; others look up to them because their thoughts and ideas move in realms beyond the imagination of those of us engaged in everyday activities. Their work brings them a step closer to creation, at least to the creation of the uninhabited world, but they are sober scientists who do not attempt to adduce ethical norms from the phenomena they observe. Their involvement with cosmic matters does not make them better human beings. They are not motivated solely by a dedication to greater knowledge. As is true of other segments of human society, thoughts of competition and career advancement enter into their calculations; quite a few discoveries grew out of just such considerations. Yet this is not to deny, as we shall learn, that we find among them a passion for knowledge and much friendly cooperation. The fruit of their research is the work of human beings and as such is often imperfect, even erroneous. But despite setbacks the course of the science of astronomy, beginning with the Babylonians and culminating in modern astrophysics, has led ever forward.

The scene has been set, the actors introduced. Let the play begin.

1

The Long Life
of Stars

The earth revolves around the sun at a rate of 30 kilometers per second in an almost circular orbit measuring 300 million kilometers in diameter. Throughout its orbit the earth is irradiated by the sun, and in turn, almost all the energy it receives is given off, particularly when rotation turns the irradiated (or day) side away from the sun, making it into the night side.

This interplay of irradiation and emission is responsible for maintaining the earth's surface at a temperature that makes our planet inhabitable. Not all the energy received from the sun is, however, given off; some of it is chemically locked into plants. Man and beast live off the solar energy stored in plants. When we burn coal or oil we are using energy stored in them from the early days of the earth's existence. The turbines of our hydroelectric plants are also powered by solar energy, for the sun's radiation evaporates the oceans, and the rains that result feed the rivers. Every square meter of the earth's surface facing the sun is irradiated by 1.36 kilowatts of power. As great as this amount of energy may seem to us, it is minuscule compared with the total energy radiated by the sun every second in every direction. If we were to express that energy in kilowatts, we would arrive at a 24-digit figure. Only a minute portion of it is absorbed by the earth.

What Is the Source of Solar Energy?

With enormous power the sun sends light and heat, and thus energy, into the cosmos year in and year out. Since when, and for how much longer? Will this radiation diminish in the course of time, thereby causing life on Earth to congeal in a blanket of ice? Or will it grow more and more intense and put an end to life by causing the oceans to boil? Never in the history of our observation of the sun has it been possible, even with the most sophisticated measuring devices, to find any deviation in the intensity of solar radiation. The assumption that the sun has been shining with undiminished strength for a long, long time is borne out by traces of organic life found in prehistoric layers of the earth's crust. Relatively highly developed unicellular organisms as complex as blue-green algae have been found in the silicified rocks of the Onverwacht stage in the Transvaal, South Africa, lending credence to the assertion that life on Earth already existed 3.5 billion years ago. The sun must have shone as powerfully then as it does today to have allowed life to exist on Earth.

The sun, being a finite entity consisting of a finite amount of matter, cannot contain an infinite amount of energy. We are able to determine the mass of the sun because its gravitational force is indicative of it. Together with other planets the earth revolves around the sun, kept in its orbit by the gravitational force of the solar mass, in which the centrifugal force at all times balances the gravitational force. This allows us to calculate the gravitational force of the sun and with it its mass (see appendix C). Expressed in metric tons it makes for a 28-digit figure. If we relate the energy produced by the sun to a gram of solar matter, we arrive at an annual yield per gram of about 5 joules. This may not seem like much when compared with the heat per gram produced by the human body, which is one thousand times greater. However, human beings must take nourishment periodically and otherwise compensate for this expenditure of ener-

9

gy, whereas solar matter has been meeting its energy needs by itself for billions of years.

What is the source of the energy that the sun has been emitting so powerfully for such a long period of time? Can chemical changes be responsible? Let us look at the simplest chemical process for the production of energy: combustion. If the sun consisted of bituminous coal, the energy produced by combustion would suffice to sustain its radiation for about five thousand years. Thus, if coal were the solar fuel the sun's fires would have gone out long ago. As with combustion so it is with all other chemical processes. They are simply not productive enough to be the sun's source of energy.

Toward the end of the last century, repeated efforts were made to find the source of solar energy. If chemical processes within the sun's interior had to be ruled out, perhaps the sun was being fueled from the outside. Our solar system is host to countless small, solid bodies called *meteors* that move among the planets. We know them from the phenomenon of *falling stars,* which appear in the sky when meteors move into the earth's atmosphere and, heating up, burn out. Some meteors fail to vaporize completely in the atmosphere; they fall to the earth, and we can see them displayed in planetariums as meteorites. Given its enormous gravitational force the solar mass is bound to attract the meteors that travel through the solar system and ram into the sun at high velocity. At the point of impact the energy of their motion would have to be transformed into heat. Can the heat thus created possibly account for solar radiation? A gram of meteoric mass colliding with the sun would produce about 190 million joules of energy. Thus, meteoric bodies whose total mass was about a hundredth part of the earth's would have to land on the sun annually to account for its radiation. Such an increase in solar mass would not have gone unnoticed, for it would have altered the earth's orbital path around the sun, and consequently, the calendar year would have had to become noticeably shorter with the passing years. But we know from accounts of solar and lunar eclipses dating back to antiquity that no substantial change in the relative movements of our planetary system has taken place, and so the hypothesis that meteors fuel the sun by ramming into it had to be discarded.

Other astronomers hypothesized that the sun's own gravitational force was the source of its energy. The German physicist and physician Hermann von Helmholtz raised this possibility in the nineteenth century. If, however, the sun were to feed off its own energy without replenishment, it would shrink in the course of time. Its diameter would grow smaller, and every gram of solar matter would—like a slow-motion fall—slowly come closer to the center of the sun. As in the fall of a meteor, energy would be released, but contrary to the process described in the meteor hypothesis, solar matter would collapse into itself. The solar mass and its gravitational effect on the earth would remain unchanged. This process could provide the sun's energy needs for about 10 million years, only a hundredth part of the billions of years that the sun has been shining. Thus, the sun's own gravitational force also had to be ruled out as the source of its energy.

Atomic Energy from the Sun and the Stars

Nuclear energy is the richest, most productive energy source we know today. A part of the electricity we use is already being supplied by nuclear power plants in which heavy uranium atoms are split into lighter atomic particles. This nuclear fission releases energy. Nuclear power plants will operate still more productively once we develop a satisfactory process for obtaining energy through the fusion of lighter atomic nuclei into heavy atoms. The fusion of hydrogen nuclei has proved to be particularly productive.

The sun, like most of the stars, is composed largely of hydrogen. Is it not therefore logical to ask whether the sun might derive its energy from the nuclear fusion of its hydrogen? As we shall see, nuclear fusion is indeed the source of solar energy. In chapter 3 we shall discuss in greater detail the nuclear processes that take place within stars. But before we proceed to demonstrate that nuclear reactions do in fact keep the sun—and hence us—alive, let us consider the consequence of the fusion of hydrogen atoms into helium atoms in the sun and in the stars, and the fact that the nuclear energy thereby released makes the sun and stars shine.

When the atomic nuclei of a gram of hydrogen fuse to produce helium nuclei, 630 billion joules of energy are released 20 million times more energy than is produced by the combustion of the same amount of bituminous coal. If bituminous coal could sustain the sun's radiation for 5,000 years and nuclear energy for 20 million times longer, the lifetime of the sun can be calculated at about 100 billion years. Finally, we have discovered a source of energy adequate to meet the demands of solar radiation for billions of years: nuclear energy, released by the change of hydrogen into helium. According to our estimate the nuclear energy stored in the sun's hydrogen will last for 100 billion years. This estimate is in fact too optimistic, for only 70 percent of the sun is composed of hydrogen, and hence its supply of nuclear fuel is a little lower than we had assumed. Furthermore, as we shall see later, the depletion of a star's nuclear energy starts to make itself felt when 10 to 20 percent of its hydrogen has been used up. We thus arrive at a time span of about 7 billion years—a period still long enough for the sun to have shone for a far longer time than life has existed on Earth.

The sun is a star like the seven thousand stars visible to the naked eye and the far more numerous fixed stars visible through a telescope. With a few exceptions they too consist mainly of hydrogen. If they too obtain their brightness from the change of hydrogen into helium, then we can also calculate how much longer their nuclear energy supply will last. For the sun that figure is 7 billion years. But other stars, like Spica, the most luminous star in the Virgo constellation, will deplete their energy supply much sooner. Because a companion star orbits Spica we are able to determine Spica's mass (see appendix C). Spica's stellar matter is about 10 times that of the sun, therefore 10 times more hydrogen is available to it. We also know that it is 10,000 times brighter than the sun. Because it radiates such an enormous amount of energy, its hydrogen supply will last only a thousandth as long as that of the sun. Thus, Spica will shine for only a few million years longer. In cosmic terms that is a very short time. After all, a million years ago higher mammals already inhabited the earth; *Pithecanthropus erectus* roamed the forests of Java.*

Stars Age

Although the amount of nuclear energy stored in the sun and other stars is considerable, in the course of time even that large a reserve becomes depleted. Stars too must age. Are we eyewitnesses to the life story of the stars? Are we able to watch as a star uses up its store of energy over time and is extinguished? As the examples of the sun and Spica have shown, changes, when measured against the span of human life, are slow. In fact, the properties of the visible stars are identical to those recorded by the Greek astronomer Hipparchus in 150 B.C. Because our scientific exploration of the stars is so recent an undertaking, we have no firsthand records of the developmental processes of stars over time. Although some stars show temporal fluctuations in brightness, these fluctuations are not directly related to long-term developmental effects. They are comparable to the flickering of a candle, which is not directly related to the depletion of the energy stored in the paraffin as the candle burns down. We have no direct observations of the effect of aging on the stars. Were we able to stay around long enough, however, we would see the stars age.

The role of the astronomer who seeks to understand the laws governing the temporal development of the stars is comparable to that of a short-lived fruit fly that would like to observe the aging process in humans. That fly, watching a person throughout its own one-day life span, will not be able to detect any substantial signs of aging.

*I have often used the example of *Pithecanthropus erectus,* Java man, in my lectures. On one such occasion a reporter for a German daily came up to me and said that he would like to run a picture along with his story on the lecture and asked where he could get a picture of Java man. I pointed out that since my subject had been the stars and *Pithecanthropus erectus* was merely a casual reference, it might be misleading if the only picture he planned to feature was of Java man. "I understand," he told me and after giving the matter some thought, added, "then we'll just have to feature a picture of you."

13

People obviously age far more slowly than flies. The fly will also be able to observe many different types of human beings—females and males, short and tall, light-skinned and dark-skinned—without knowing whether it is seeing different types or different age groups of one and the same type. It will receive only a momentary impression. It will not and cannot know whether short persons will always remain short, whether light-skinned persons will perhaps turn dark, whether males will turn into females. We are in a similar position with regard to the stars. We receive a momentary impression of the totality of all the stars, and there exists a vast number of different and sometimes puzzling kinds. For example, a strange star moves around Sirius.

The Companion of Sirius

Sirius is the brightest fixed star of the night sky. In 1844 Friedrich Wilhelm Bessel, the director of the Königsberg Observatory, noticed that Sirius periodically executed a barely perceptible maneuver (figure 1.1). He concluded that Sirius had to have a satellite and that together they moved around a common center of gravity in a fifty-year period. His assumption was met with doubt because no satellite was visible. In 1862 Alvan George Clark, a renowned American lensmaker, discovered a small, barely visible star very near Sirius—Bessel's satellite—while testing the optical system of a refracting telescope he had built for a Chicago observatory.

Today we know somewhat more about the two stars. Every 49.9 years they complete an orbit around a common center of gravity. The study of the motion of this binary system has yielded information about them. The primary star—Sirius A—has 2.3 times the mass of the sun. The second, long-unseen star—Sirius B—has less mass, about the same as that of the sun. The two stars are completely different. Sirius A is twice the size of the sun. An average cubic centi-

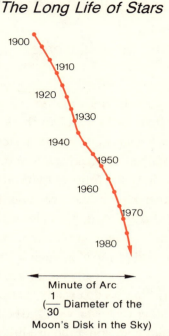

1900
1910
1920
1930
1940
1950
1960
1970
1980

Minute of Arc

$(\frac{1}{30}$ Diameter of the

Moon's Disk in the Sky)

Figure 1.1 The path of Sirius in the sky from 1900 to 1985. Like all so-called *fixed stars,* Sirius too moves slowly through the sky. Here the movement is shown going from top left to bottom right. The motion is caused by the somewhat different orbits of Sirius and of the sun around the galactic center. The uniformity of Sirius's orbit is disturbed every fifty years, shown here by a notch at 1940. As can be seen from the scale at the bottom of the figure, both the uniform motion and the disturbances change Sirius's position in the sky only imperceptibly. The movement is detectable only with the most delicate measuring devices. The regularly recurring disturbances are caused by a faint companion revolving around the primary star coming close to it at fifty-year intervals, thereby interfering with its uniform course.

---

meter there contains about a quarter gram of matter, somewhat less than does a cubic centimeter of the sun's mass, which contains about a gram. Not so Sirius B. Its radius measures only about one-hundredth that of the sun, and since its mass is the same as the sun's, it must be a million times as dense. Presumably, a cubic centimeter of Sirius B contains 1,000 kilograms of matter. The Sirius system thus joins together two completely different stars. Stars like Sirius B are not uncommon and also appear singly. Most of them have high surface temperatures and give off a white light. Because of their small size they are known as *white dwarfs.*

The Supergiant in the Charioteer

White dwarfs are stars in which matter is compressed one million times more densely than in the sun. But there also exist stars of far less density. As with Sirius it is a binary system that now offers us the opportunity to study an interesting star of extremely low density.

Whenever the gravitational force of two stars forces them into an orbit around each other, astronomers consider themselves lucky, for the relative motion of these stars furnishes a clue to the amount of matter that creates the common gravitational field. It is particularly fortuitous if the two bodies move in such a way that, viewed from our solar system, one is periodically positioned behind the other and hence eclipsed. Many binaries do exactly that. In them the two stars are so close to each other that even with the best telescope their light is seen as a single point combining the brightness of both. But when one star moves in front of the other and covers it, they appear less bright. Their combined radiance appears diminished until the eclipsing star moves away again and its companion reappears. Such binaries are known as *eclipsing variables* because their brightness changes.

The way in which their brightness waxes and wanes and the difference between the successive eclipses, when eclipsed and eclipsing stars reverse positions, supply us with information about the properties of stars. I mention this now merely because in the 1930s an eclipsing star enabled us to examine a special type of star, a *supergiant,* more precisely than we had ever thought possible. The star system in question was Zeta Aurigae in the Charioteer constellation. Astronomers had long known that Zeta Aurigae was a binary system even though it did not, unlike Sirius, look like one when seen through a telescope. But closer examination had revealed that its light appeared to emanate from two stars, a hotter and a cooler one. This discovery had led to the conclusion that it was indeed a binary system, and astronomers wondered if the two stars were perhaps eclipsing variables.

In the winter of 1931–32 the German astronomers Heribert Sch-

16

The Long Life of Stars

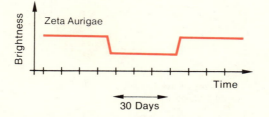

Figure 1.2 The brightness curve of Zeta Aurigae. Within a 24-hour period its brightness diminishes by 65 percent. The star remains at this fainter level for 37 days and then regains its normal brightness within 24 hours. After 972 days this cycle starts anew.

neller and Josef Hopmann did indeed discover the eclipse by using photometers, precise instruments for measuring stellar brightness. Within a period of 24 hours the radiance of the two stars diminished by 65 percent, remained at this fainter level for 37 days, and then within 24 hours regained its normal value. This process repeated itself every 972 days (figure 1.2).

The study of this phenomenon during subsequent eclipses yielded a wealth of information about the system. These, briefly, were the major findings: the hot star Zeta Aurigae B has a surface temperature of about 11,000°, is approximately 3 times the size of the sun, and has about 10 times its mass. The cooler star Zeta Aurigae A has a surface temperature of only about 3,400°, compared with the sun's surface temperature of about 5,800°.* Zeta Aurigae A is composed of 22 times more stellar matter than the sun, and its radius—and that's what's so exciting—is 200 times that of the sun. This star is so gigantic that it could easily accommodate not only the sun but the orbit that the earth describes around the sun annually. The observed brightness minimum occurs when the hot star disappears behind the cool giant star and remains eclipsed for thirty-seven days. But when it moves in front of the cooler star, it covers only a small part of that star's surface (figure 1.3). Because the eclipsed portion

*Here and subsequently, unless otherwise stated, we use the physicists' absolute temperature scale, which has its zero point at −273° C. A temperature difference of 1° is the same as that of 1° C. We therefore arrive at the Celsius temperature if we subtract 273°. The surface temperature of the sun is therefore about 5,530° C.

17

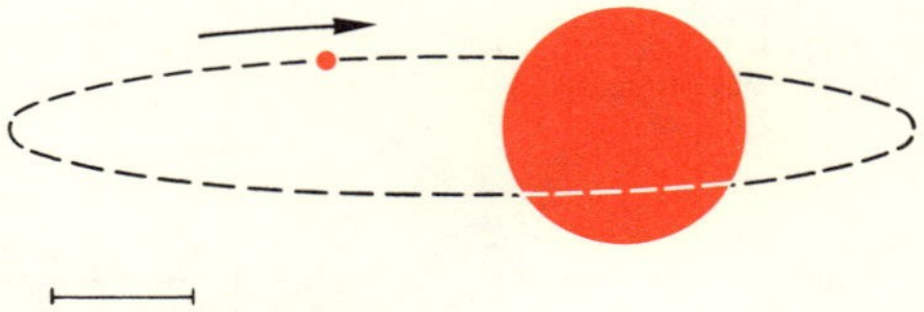

Diameter of the Earth's
Orbit around the Sun

Figure 1.3 The Zeta Aurigae binary as it would look from the earth if one were able to see its details by telescope. In fact, however, the stars are not seen separately; their light blends into a single spot of light. When the smaller star, which contributes more than half of the system's brightness, is eclipsed by its bigger companion, we receive light only from the latter. The observed brightness of the system is then fainter (see figure 1.2). The small star takes 972 days to complete its orbit around its big companion.

of the big star does not contribute substantially to the radiance of the system, the second eclipse is not perceptible.

In the Zeta Aurigae system we have learned something about two specific stars. Its hot star is not too unlike the sun and Sirius A. True, it has more mass and its diameter is greater, but its mean density, the amount of matter contained in a cubic centimeter, is still comparatively close to the sun's: 1/3 gram of matter per cubic centimeter. The cool star is an entirely different matter. There we find an average of only 3 millionth gram of matter per cubic centimeter. Stars of this type are known as supergiants.

By now we have got to know three substantially different types of stars. Normal stars—let us call them that for the time being—like the sun, Sirius A, and the hot component of Zeta Aurigae are stars with medium densities that range from a few tenths to a few grams of matter per cubic centimeter. *White dwarfs* have extremely high densities averaging 1,000 kilograms per cubic centimeter. Finally, there are *supergiant stars,* which have densities of a few millionths of a gram per cubic centimeter. One also knows of stars, big compared to the sun but not as big as supergiants. They are called *giant stars.*

Although stars of these types show up only as tiny dots of light even in the most powerful telescopes—dots that differ merely in color and brightness—even the sort of superficial study of these objects we have carried out above gives an inkling of the multitude of phenomena the world of stars has to offer. To understand this vast variety we must try to bring order into the 100 billion stars that together with the sun fill our galaxy.

2

The Astrophysicist's Most Important Diagram

In the previous chapter we learned the various forms that stars can take. There are bright blue stars with high mass and dim reddish ones with low mass. We know of the big red giants and supergiants and of the small white dwarfs, and here on our planet Earth we are like short-lived fruit flies as we try to figure out the temporal sequence governing this array.

The problem facing us might seem overwhelming, yet it has been solved. We now know the path of development the stars follow, at least in its essentials. To understand how this development has been determined, it is first necessary to establish order among the variety of stars by classifying all visible stellar bodies according to measurable criteria.

Measuring and Classifying the Stars

Surface temperature, the simplest quantity to measure, lends itself most readily to our purpose. It can be ascertained with comparative ease by relying on the color of the star. When looking at the stars

20

in the sky, most of us do not realize that they differ in color, but when we look at pictures taken with color filters, these color differences become obvious. Blue stars are hot, red stars comparatively cool. However, a star's color gives only an approximation of its temperature; its spectrum is a more reliable indicator thereof (see appendix A). It is possible to determine the temperature of the luminous surface of any star that shines brightly enough. For example, Sirius A, the primary star in the Sirius system and one of the hotter stellar bodies, has a surface temperature of 9,500°. In the Orion nebula we find stars with surface temperatures as high as 20,000°. On the other hand Betelgeuse, the brightest star in Orion, is of a reddish hue, which makes it a cool star; its surface temperature is 3,000°. As mentioned earlier the sun has a surface temperature of 5,800°.

Another telling attribute of a star is its *luminosity,* that is, the energy it radiates per second. This quantity cannot be determined through direct observation. True, we can measure how bright a star appears in the sky, but that does not tell us the amount of energy it emits. Stars of equal luminosity can appear to differ in brightness if they are not equidistant from us. Because of the way light propagates, a more distant star appears fainter than an equally luminous one closer to us. To find out the amount of energy radiated by a star from its brightness, we must first know how far away it is. (Appendix B describes the methods used by astronomers to arrive at that distance.) In other words we can only determine the luminosity of a star whose distance is measurable. To us the sun appears to be the brightest fixed star in the sky, yet there are others of far greater luminosities, the most powerful of which radiate energy of a rate 100,000 times greater than that of the sun. Yet because they are so far away, they look like rather insignificant smudges in the sky. Conversely, there are some puny stars whose luminosity is 1/100,000th that of the sun.

We have now been introduced to two important measurable properties of stars, surface temperature and luminosity, and can therefore consider an interesting question, namely, whether these two properties can and do appear in any number of possible combinations. For

21

instance, do we know of stars with a high luminosity that are hot and others with the same power that are cool? Is it possible for both hot and cool stars to have very low levels of luminosity?

The Hertzsprung-Russell Diagram

Astronomers found the answer to these questions in a diagram correlating the stellar surface temperatures of a great number of stars with their luminosities. This diagram has proved of inestimable help in our effort to decipher the code of stellar evolution. Invented by the Danish astronomer Ejnar Hertzsprung and his American colleague Henry Norris Russell, the diagram that bears their names is, for the sake of brevity, referred to as the H-R diagram. The temperature of a star as deduced from its color is recorded along the diagram's horizontal axis from right to left (see figure 2.1). If we also know the star's distance, we are able, based on its brightness in the sky, to determine its luminosity, which is recorded along a vertical axis from the bottom up. On figure 2.1 we have entered these two attributes for some stars mentioned earlier. At the 1,000 level on the vertical axis are recorded stars whose luminosity is 1,000 times that of the sun. The sun is thus vertically located at the level of 1, and given its surface temperature of 5,800°, it stands at the center of the diagram. Stars of greater luminosity are above it, and less luminous ones like Sirius B, the white dwarf in the Sirius system, are below it. Stars that are hotter than the sun, such as Sirius A; Zeta Aurigae B, the hot star of Zeta Aurigae; and Spica are found to the left of the dot connoting the sun. Cooler stars like Betelgeuse and the supergiant in Zeta Aurigae are to its right.

The dots on the H-R diagram tell us something about the properties of the stars. The cool stars give off a reddish glow. We look for them on the right side of the diagram, whereas hot stars, which give off a white or blue light, stand on the left. The luminous stars are

The Astrophysicist's Most Important Diagram

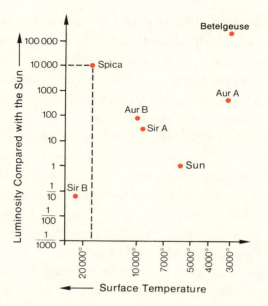

Figure 2.1 The Hertzsprung-Russell diagram showing some of the stars we already know. Once we know the surface temperature of a star, we can draw a vertical line at that particular temperature on the horizontal scale. And if we also know the star's luminosity, we can draw a horizontal line at that point on the vertical scale. We can then enter the star at the point at which these two lines intersect, as shown for Spica (surface temperature, 18,000°; 10,000 solar luminosities). The abbreviations *Sir* and *Aur* in the diagram stand for the components of the Sirius and Zeta Aurigae binaries discussed in the text. (The temperature scale on the horizontal axis refers to absolute temperatures as explained in the footnote on page 17.)

---

in the upper portion; those with less radiated power, in the lower. Thus, cool stars of great luminosity stand in the upper right-hand side of the diagram. One square centimeter of the surface of a cool star does not emit a great deal of energy per second; yet since the star as a whole sends out a great deal of energy, its surface must be big enough to accommodate a great many square centimeters. In other words it has to be big, which means that all the stars of the upper right-hand side are big. They are known as *red giants* and *red supergiants*. This fact confirms something we alluded to earlier, namely, that the primary star of Zeta Auriage is huge; it could accommodate the orbit of the planet Earth.

Let us now take a look at the lower portion of the diagram. There we find hot stars of little luminosity. Since one square centimeter of such a star's surface emits a great deal of energy per second though the star as a whole does not, it stands to reason that the star is small. The white dwarfs, among them Sirius B, are thus located in the lower left-hand portion of the diagram.

In general, the luminosity and surface temperature of a star give a clue to its size. Its temperature tells us the amount of radiation per square centimeter of surface, whereas the total radiation as indicated by its luminosity tells us the size of its radiating surface and with it the star's radius.

Before showing how the H-R diagram helps us understand the evolution of the stars, one brief observation is in order: the radiation emitted by a star is not easy to measure, for the earth's atmosphere constitutes a barrier. Shortwave light, as for example ultraviolet radiation, does not make it all the way down to us. But even radiation that does reach us is not easily measured. The human eye can detect only a portion of solar and stellar radiation, and photographic film cannot capture all of it. Eye and film perceive different color intensities of light differently. When we speak of a star's luminosity, we generally mean the radiation discernible to the human eye. To measure it we use instruments that, with the help of filters, take into account the color sensibility of the human eye. The H-R diagram usually indicates a star's *visual luminosity* rather than its *total luminosity* which takes into account the radiation of all frequencies.* This, however, does not substantially affect the H-R diagram. Our diagrams here, when based on observed data, mainly indicate visual luminosity, but when computer data are used they indicate total luminosity. Each diagram clearly spells out the type of data it is based on.

---

*The distinction between total and visual luminosity is not merely one of semantics. For instance, a star of 10 solar masses, such as Spica, emits 10,000 times as much energy as the sun, but visually it appears to be only 1,000 times as bright.

Stars near the Sun

We now have all the information we need for the H-R diagram. Let us begin with the stars in the neighborhood of the sun. By that we mean stars "close" to us, that is, stars no more than 70 light-years away. (This means it takes their light less than seventy years to reach the earth.) These stars are indeed close, considering that it takes the light from the most remote stars of the Milky Way seventy thousand years to show up in the telescopes of astronomers. We are now receiving light and radio waves sent out by remote galaxies billions of years ago, at a time when the cosmos was still in its infancy. The stars we are about to discuss are nearby, even though they are much farther out in space than the sun. It takes light only about eight minutes to travel from the sun to the earth, while it takes the light from Proxima Centauri, the second closest fixed star, four and a half years to reach our planet.

What makes the nearby stars so interesting is that their distance can be determined fairly precisely (see appendix B) because it helps us to go from brightness to luminosity. The distance of the stars plus their brightness in the sky enables us to determine their luminosity, or more precisely, their visual luminosity as determined from their distance and their visual brightness in the sky, which is determined by a photometric device equipped with a color filter to measure visual radiation.

We find the surface temperature by yet another measurement with yet another color filter, generally a blue one. By noting a star's brightness in the blue light and in the visual range, which tends to be of a somewhat reddish hue, we are able to determine its color and hence its surface temperature. Once we know both surface temperature and visual luminosity, we can enter the star's position at the appropriate spot on the H-R diagram. Figure 2.2 shows the positions of the stars near the sun. What jumps out at us when we look at this diagram is the uneven pattern of the dots on the plane. Most of them lie along a strip winding downward from left to right, that is, from

25

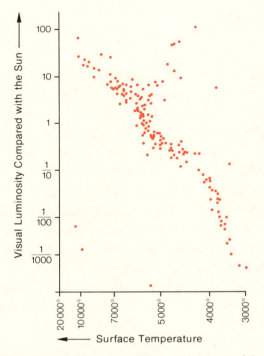

Figure 2.2 The H-R diagram of the stars in the neighborhood of the sun. The temperature and luminosity of most of them is such that the dots designating them in the diagram lie along a strip winding downward from left to right. This strip is called the *main sequence*. Some of the stars are found on the upper right, and they are known as *red giants*. Some are found on the lower left, and these are called *white dwarfs*. (Adapted from R. H. Stoy)

bright blue to faint red stars. Only a few of the stars at the right lie in the area of the red giants, and on the lower left we find only three white dwarfs.

Ninety percent of all stars lie along the aforementioned strip, and for that reason astronomers call it the *main sequence* and the stars themselves *main-sequence stars*. Looking at figure 2.1 we can see that the sun, Sirius, and Spica inhabit the main sequence, but the cool star in Zeta Aurigae or Betelgeuse or the companion star of Sirius do not. Main-sequence stars are, so to speak, the average citizens in the solar neighborhood, whereas the giants and dwarfs are among the exceptions.

The main-sequence stars display another important property, one linked to their mass. We know the amount of matter possessed by a star in only a few isolated instances, for only when a companion star revolves in the gravitational field of a star can the latter's mass be determined with any degree of accuracy. Thus, the movement of the planets has enabled us to calculate the mass of the sun, and through the Sirius companion we have learned that Sirius A is composed of about 2.3 times the amount of the solar mass and the companion itself of about 1 solar mass. Through the motion of a companion we have been able to determine the mass of a number of other stars (The method is outlined in appendix C.). The most massive stars we know are composed of 30 to 50 times the solar mass and the least massive stars of only 1/10th the solar mass.

Looking at the main-sequence stars whose masses were established through their companion stars, we find the following surprising fact: specific segments of the main sequence contain only stars of a specific mass (figure 2.3). Low-mass main-sequence stars are at the lower

Figure 2.3 H-R diagram with main sequence (the red line in the figure). Only stars of a specific mass are found at any one point of the main sequence. (Astronomers generally use solar mass as the standard unit and $M_\odot$ as the symbol for it.)

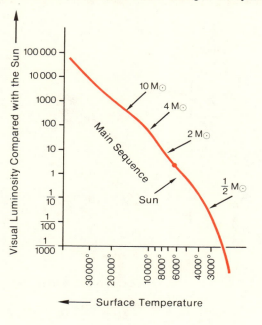

end, massive ones at the top. If we go up along the main sequence from bottom to top, the stars gradually increase in mass. Because luminosity in the H-R diagram also increases from bottom to top, it is safe to say that with increasing luminosity the mass of main-sequence stars also increases. If we know which of two main-sequence stars has greater luminosity, we also know which is the more massive. We can even go further and say that luminosity is a direct indication of mass, provided that we are dealing with main-sequence stars. Figure 2.4 shows the parallel increase of luminosity and mass in main-sequence stars. The astronomer calls this the *mass-luminosity relationship*. The sun, Sirius A, and Spica demonstrate this relationship particularly well, whereas Sirius B, not being a main-sequence star, does not follow this pattern.

We have now classified the visible stars in the neighborhood of the sun and found that (1) the H-R diagram establishes the main

---

Figure 2.4 If the vertical left axis of a diagram is used to indicate luminosity and the horizontal axis stellar mass, the main-sequence stars will lie along a narrow strip: the greater the stellar mass, the greater the luminosity. This is known as the *mass-luminosity relationship*. In the diagram the luminosity of Sirius B, the Sirius companion, is smaller than the luminosity of a main-sequence star of like mass. Sirius B does not satisfy the mass-luminosity relationship.

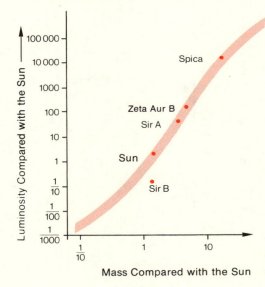

sequence; and (2) the mass-luminosity relationship holds true for main-sequence stars. But what, you may ask, does this have to do with the development of the stars? This question brings us back to the fruit fly analogy. We look at stars possessing different characteristics, just as the fruit fly sees human beings who differ in appearance. The main sequence shows us an array of external features, but we don't know what to make of it. We are like that fly, which classifies people according to externals, perhaps by the shape of their ears, yet has learned nothing about how human beings develop.

We who know something about the process of aging in humans could give the fly some helpful hints. We might tell it, for example, that classes in schools are composed of persons of like age. Knowing this, the fly could figure out that sex and color were not developmental attributes, that persons of different sex and different colors are not one and the same individual at different stages of life. The fly would also notice that body size is closely linked to age. The astronomer is in the fortunate position of finding "school classes" of stars in the sky, that is to say, groups of stars composed of individual members of like age.

Star Clusters—"School Classes" of Stars

At times the stars develop a kind of herd instinct. They crowd together and form groups, called *star clusters*. Some of these clusters were already known to the ancient Greeks and Romans. Their poets celebrated the Pleiades, a cluster named for the the seven daughters of Atlas (Plate III). The brightest can be seen with the naked eye. As a matter of fact a great many more weaker ones, at least 120 and probably as many as several hundred, form this cluster. The stars of the Pleiades are contained in a comparatively small space; it takes light thirty years to traverse the cluster from one side to the other. That the Pleiades do indeed constitute a cluster is underscored by the fact that in comparison only about twenty stars populate a sphere around the sun that is 30 light-years in diameter. Not only are the Pleiades crowded into a small area but they also move at the same rate in the same direction. These two factors—occupying one small area and moving at the same velocity—lead us to believe that the stars of the Pleiades were born at the same time. This assumption applies to other clusters, such as the Hyades, which were also known to the ancient Greeks and Romans. And the common origin of the stars in the so-called *globular clusters* (see figure 2.5), composed of between 50,000 and 50 million stars, is even more obvious. The stars in the dense center of these clusters are 10,000 times closer together than the stars in the neighborhood of the sun. What a fantastic sight the starry sky must offer to an inhabitant of a planetary system whose sun belongs to a globular cluster!

What is the luminosity-surface temperature relationship of the stars in a cluster? Is it similar to that of the stars near the sun, as seen in figure 2.2? Are most of the stars in clusters main-sequence stars? When we look at their H-R diagrams, we find substantial differences. Although all the stars of some clusters lie along the main sequence, as we can see from the H-R diagram of the Pleiades (figure 2.6), in the majority of clusters only the weaker, less bright stars possess main-sequence characteristics and the course of the main sequence is incomplete. The sequence breaks off as we move toward greater intrinsic brightness: the brighter main-sequence stars are missing. Instead, we find that the clusters contain bright red stars—red giants and supergiants—as shown in the H-R diagram of the Hyades (figure 2.7) and seen even more clearly in the H-R diagram of a globular cluster in figure 2.8. Here only the lower segment of the main sequence is populated, whereas the dots indicating the brighter stars are almost invariably farther to the right. This pattern becomes even more apparent if we plot the stars of various clusters on a single H-R diagram, as shown in figure 2.9. In this figure the main sequence is indicated by a thin line, and broad pink, broken lines mark the positions of the stars of the different clusters. All clusters share a part of the main sequence, but in the higher luminosities they veer to the right. Also, the point at which the line indicating

Figure 2.5 The globular cluster 47 Tucanae photographed with the 1-meter Schmidt mirror of the European Southern Observatory (ESO), Chile. The density of the stars in the cluster is so great that in the photograph the center does not look like a collection of individual stars. The picture gives the impression that the stars in the center of the cluster are touching, whereas in fact they are at a considerable distance from each other. (Photograph: ESO-European Southern Observatory, Munich)

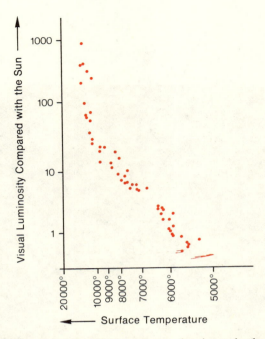

Figure 2.6 H-R diagram of the Pleiades cluster showing only the brightest of its stars. The stars exhibit a sharply defined main sequence. However, toward the top, at visual luminosities more than 1,000 times that of the sun, the stars in the diagram diverge from the main sequence somewhat toward the right. (Adapted from H. L. Johnson and W. W. Morgan)

Figure 2.7 H-R diagram of the Hyades cluster. Whereas the Pleiades (see figure 2.6) contain main-sequence stars of up to 1,000 solar luminosities, the main sequence of the Hyades ends at somewhat less than 100 solar luminosities. The brighter main-sequence stars are missing. Instead, the H-R diagram of the Hyades contains a number of red giants. (Adapted from H. L. Johnson)

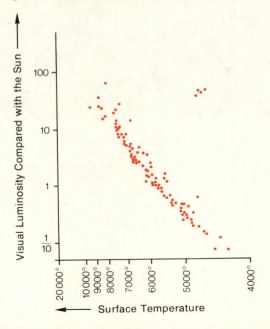

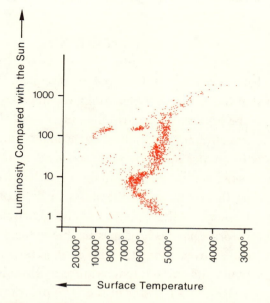

Figure 2.8 H-R diagram of M3, a globular cluster similar to the one in figure 2.5. Only stars of 5 solar luminosities are still found on the main sequence. Most of the more luminous ones are not main-sequence stars. Later sections of this book discuss stars of 100 solar luminosities lying along a horizontal strip with temperatures ranging from about 5,800° to 13,000°. That strip is known as the *horizontal branch*. (Adapted from H. L. Johnson and A. R. Sandage)

Figure 2.9 The departure of various star clusters in the H-R diagram from the main sequence (adapted from A. R. Sandage). The heavy broken lines show the position of the stars of different clusters. The main sequence of a cluster in Perseus is the topmost one, and then branches off toward the right. The globular cluster M3 has the shortest main sequence, branching off to the right at a much lower luminosity. The arrows on the left indicate the position of main-sequence stars of a specific mass. The figures next to the arrows give star masses in solar-mass units. Thus, the cluster in Perseus contains main-sequence stars of 10 to 15 solar masses, whereas the most massive main-sequence stars of M3 contain only 1.3 solar masses.

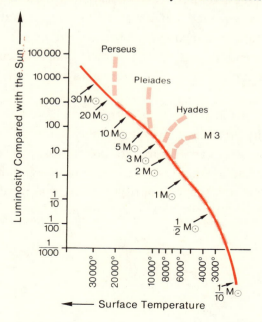

a particular cluster branches off from the main sequence differs from cluster to cluster. Since we know that along the main sequence mass increases from the bottom up, we can also say that within a cluster stars below a certain mass are main-sequence stars but that in the area of stars with greater mass the main sequence is not populated. It is this conclusion that in the final analysis has handed us the key to understanding the evolution of the stars. In this connection the work of the astronomer Allen Sandage of the Mt. Wilson and Mt. Palomar observatories has been of inestimable value.

As a star evolves over time (that is, as it ages), its properties change, in particular its surface temperature and luminosity. As it changes the dot signifying the star in the H-R diagram moves. For example, if a red giant in the course of millions of years were to turn into a white dwarf, its dot in the H-R diagram would move from the upper right to the lower left. If we lived long enough we would be able to measure the stars over millions or billions of years, record the findings on the H-R diagram, and watch the dots change their positions, moving rapidly through some areas and resting for longer periods in others. We would be able to follow the life of the stars on the H-R diagram. What we get instead is only a moment in time. All we see is where the stars are today or more precisely where they were when they sent forth the light we are receiving.* What strikes us here is that the stars in the neighborhood of the sun are clustered along the main sequence. Does that perhaps mean that the dots in the H-R diagram pass through the band of the main sequence only very slowly, spending a great deal of their time in transit there? In that case a great many stars of different ages would have to be found there.

This phenomenon parallels something we have observed in our own lives. Why are there more adults than children? The answer is simple: human beings are children for only fifteen years of their lives and then spend an average of fifty or more years as adults. Take any group of people in a town comprising different age groups and the

---

*Because the time it takes for the light of the stars in the Milky Way to reach us is relatively short compared with the time over which stars evolve, this difference does not matter too greatly.

majority are likely to be adults. Is the main-sequence phase perhaps one in which a star spends a considerable period of time?

We know that the sun is a main-sequence star, and we also know that it has changed hardly at all in billions of years; it has been a main-sequence star for all that time. As we learned, the energy contained in the sun's hydrogen has been able to cover its radiation for that period. Is it perhaps possible that all main-sequence stars meet their energy needs through hydrogen fusion? Since hydrogen is such a rich source of energy, is the reason the stars in the H-R diagram remain unchanged for such long periods perhaps also why they cluster along the main sequence?

Let us assume that all main sequence stars meet their expenditure of energy by transforming their hydrogen into helium. From our earlier estimates concerning the sun and Spica, we know how long the stars can radiate with their hydrogen fuel. Assuming that 70 percent of the mass of a star consists of hydrogen and that the depletion of nuclear fuel makes itself felt when 10 percent of the hydrogen is transformed, we arrived at a life span of 7 billion years for the sun, whereas Spica, with 10 times the sun's mass and about 10,000 times its luminosity, can continue to shine with undiminished intensity for only a few million years. By the same method we can calculate how long every main-sequence star can cover its radiated energy through hydrogen fusion. Let us take any one of the stars in the main sequence of figure 2.3. We can determine its luminosity in the diagram and through the mass-luminosity relationship of main-sequence stars in figure 2.4 find the mass matching the luminosity. A comparison of the nuclear energy stored in this mass and the luminosity, that is, the energy radiated into the cosmos per second, will tell us how long this energy supply will last. The main sequence plotted in figure 2.10 shows how long stars of different mass can live on their hydrogen fuel. It confirms the assumption based on the comparison of Spica with the sun, namely, that the greater the mass of a main-sequence star, the greater its expenditure of energy, thereby shortening the duration of its hydrogen supply.

Those who devote their lives to the study of stars find many simi-

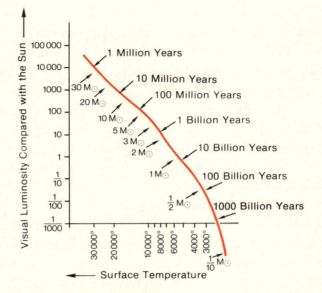

Figure 2.10 The main sequence in the H-R diagram. The arrows on the left indicate the location along the main sequence of stars of a specific mass (in solar-mass units, $M_\odot$). Since the mass determines the available nuclear energy and since the luminosity at every point of the main sequence is known, the length of time that a star at any given place along the main sequence can live from its hydrogen fuel can be calculated. These time spans are indicated by arrows to the right of the main sequence. Stars of more than 30 solar masses can subsist on their hydrogen fuel for barely a million years, whereas stars of 0.5 solar mass can live off it for almost 100 billion years. A comparison with figure 2.9 enables us to estimate the age of star clusters.

---

larities between human beings and the stars—in this instance, the greater the mass, the shorter the life expectancy.

The Age of Star Clusters

In a group of main-sequence stars of different mass but identical age, all living off hydrogen fusion, the effects of hydrogen depletion would first become noticeable at the upper end of the main sequence, the site of the more massive stars. With the passage of time the stars of increasingly smaller mass would also deplete their energy store. After 7 billion years stars of 1 solar mass would show symptoms of depletion.

But is this not precisely the process we find in star clusters? Let us take another look at the H-R diagram of the Hyades in figure 2.7. The main sequence of this cluster is crowded in the upper portion of the diagram up to a visual luminosity of about 20 times that of the sun, equivalent to stars of about 2.5 solar masses. The hydrogen life span of this type of star is about 800 million years (see figure 2.10). When a group of stars of the same age live for 800 million years off hydrogen fusion, stars of 2.5 solar masses would just begin to feel the depletion of their hydrogen, whereas only less massive stars can continue to live off their stored hydrogen fuel. Is that perhaps why the upper portion of the Hyades main sequence is not populated?

Other star clusters depart from the main sequence at different levels of luminosity and thus at different levels of mass. The Pleiades, for example, still contain main-sequence stars of 140 solar luminosities (140 times as luminous as the sun), which are equivalent to stars of a little more than 6 solar masses, and their hydrogen supply will sustain them for only 100 million years. In the H-R diagram of the Pleiades, the stars of greater luminosity are not found directly on the main sequence but lie somewhat to the right of it: the first sign of their depletion. By constructing an H-R diagram and seeing how far up the main sequence is populated, we can generally establish the ages of star clusters. Figure 2.9 is a schematic representation of a number of star clusters and their deviation from the main sequence. A star cluster in Perseus is the youngest; its main sequence is populated up to the level of 1,000 solar luminosities. The cluster is about 10 million years old. Next come the Pleiades, followed by the Hyades, and finally the aged globular star cluster M3. Its main sequence is populated up to about 3 solar luminosities, and its main-sequence stars have somewhat less than 1.3 solar masses. They appear to be in the process of departing from the main sequence, which would indicate that the cluster is about six billion to ten billion years old.

If the departure of the star clusters from the main sequence in the H-R diagram does indeed indicate that their hydrogen store is nearing depletion, we have made giant strides toward gaining understanding of the evolution of stars. For this would mean that stars remain on the main sequence until their hydrogen supply is depleted, after which they move toward the right into the region of the red giants, since the stars that have left the main sequence are found on the right. If that is in fact so, new questions arise. How old are the oldest star clusters, and how young the youngest? What were the stars before hydrogen fusion set in? What happens to a star once its hydrogen store is depleted? We know that stars turn into red giants, yet they cannot continue to shine for too long; their nuclear energy has been largely expended.

We must keep in mind that at present we are merely *conjecturing* that the properties of the stars in clusters are related to the depletion of their store of nuclear energy. Observation supports our hypothesis, yet the devices available to us cannot even tell us whether the temperature and density in a star's center are high enough for the nuclear reactions that would let the star operate like a nuclear power plant. We know that their surface temperatures certainly are not. But how can we know the temperature inside the stars? The light they emanate is emitted by a thin surface layer. In the case of the sun the light comes from an atmosphere whose mass amounts to only a hundredth of a billionth part of its total mass. We cannot look into it more deeply. Still, we know more about the interior of the sun than about the interior of our own planet Earth. Why this is so will be explained in the following chapters.

3

Stars As Nuclear Power Plants

We still do not know for certain whether nuclear reactions are the source of stellar radiation. True, we have not yet found another equally abundant source of energy, but that does not necessarily mean that none exists. Couldn't it after all be possible that future scientists, in the best tradition of science fiction, will discover hitherto unknown sources of energy?

In the previous chapter we saw that some of the properties of star clusters support the assumption that nuclear energy is produced in the stars, and as we shall see here and in subsequent chapters, the notion is valid. There is no need to search for new, yet unknown sources: nuclear physics has shown persuasively how and why stars shine. Yet not too long ago, in the early 1920s, physicists refused to entertain the possibility of nuclear reaction in the stars, which had something to do with the structure of atoms.

The Building Blocks of Atoms

All matter in the universe—rocks and minerals, air and oceans, plant and animal cells, gas nebulae and stars—ultimately is composed of only ninety-two building blocks called *chemical elements*.

This insight, a nineteenth-century breakthrough, substantially simplified our concept of matter. In this century we were able to show
that in the final analysis these ninety-two chemical elements are composed of only three types of building blocks—*protons, neutrons,* and
electrons. Helium atoms, for example, differ from carbon atoms only
in the different quantitative arrangement of these three basic building
blocks (see figure 3.1).

A helium atom consists of a nucleus containing two protons and
two neutrons. The proton is a positively charged particle, and the
neutron has no charge. Consequently, the nucleus of a helium atom
is positively charged. Around it revolve two electrons, which are neg-
atively charged light particles; they form the electron shell of the helium atom. The structure of carbon is more complex. It too consists
of a nucleus containing protons and neutrons, in this case six protons
and six neutrons, around which six electrons revolve. The simplest
atom is the hydrogen atom. Its nucleus consists of one proton,
around which one electron revolves.

Protons and neutrons are of about equal mass. They are called

Figure 3.1 Schematic representation of hydrogen, helium, and carbon atoms. Protons are in red, neutrons dark gray. The orbits of the electrons around the nuclei
are not drawn to scale. In the carbon figure the six electrons orbiting the nucleus
have been omitted.

Hydrogen
Mass Number 1
Atomic Number 1

Helium
Mass Number 4
Atomic Number 2

Carbon
Mass Number 12
Atomic Number 6

heavy particles, even though compared with our idea of what is heavy they weigh practically nothing. One trillion such heavy particles would weigh only about a trillionth of a gram. An electron has only about a 1/2,000th part of the mass of a proton. Because the proton has a positive charge and the electron a negative one, together they are electrically neutral. As we mentioned the neutron has no charge. Occasionally, we come across a *positron,* a particle of the mass of an electron with a positive charge. But the positron cannot survive for long, for if it enters the vicinity of an electron it promptly unites with it, and the two go up in a small flash of light.

All atomic nuclei are composed of a certain number of protons and neutrons. The number of electrons that normally revolve around the atom is equal to the number of protons in its nucleus, so the positive charge of the protons of the nucleus is canceled by the negative charge of the electrons. In fact, atomic structure is even simpler. Strictly speaking, matter consists not of three types of building blocks—protons, neutrons, and electrons—but only of two. In atomic nuclei protons and electrons can combine into neutrons, and a neutron outside an atomic nucleus breaks down into one proton and one electron within about seventeen minutes. Thus, it is safe to say that essentially the universe in all its diversity consists of only protons and electrons.

The number of protons plus the number of neutrons in an atomic nucleus constitute the *mass number* of the nucleus; the number of its protons is its *atomic number.* Hydrogen atoms thus have the mass number 1 and the atomic number 1. Helium's mass number is 4, and its atomic number is 2. The most common type of iron atom has the mass number 56 and the atomic number 26. The atomic number also indicates the number of electrons that have to revolve around the nucleus to make the atom electrically neutral. The electron shell determines the chemical characteristics of matter. Elements with different atomic numbers differ chemically because they have different electron shells. Atoms with the same atomic number but with a different number of neutrons are chemically identical. Known as *isotopes* of one and the same element, they differ only in their mass number. Besides the ordinary hydrogen there is a hydrogen isotope, called *heavy hydrogen,* whose nucleus contains a neutron in addition to the proton. This hydrogen isotope is also called *deuterium* and in nature occurs only in minute quantities.

Even though a piece of iron is very different from the hydrogen gas in a balloon, both are agglomerations of protons and electrons. If one were to take fifty-six hydrogen atoms and arrange their fifty-six protons and fifty-six electrons at random, then take thirty of these electrons and protons and fuse them into thirty neutrons, then combine the neutrons with the remaining twenty-six protons into an atomic nucleus, and finally let the remaining twenty-six electrons revolve around this nucleus, one would have created an iron atom out of hydrogen. Similarly, if one were to take four hydrogen atoms and out of two electrons and two protons make two neutrons and combine them with the remaining two protons into an atomic nucleus, one would get an atomic nucleus with the mass number 4 and the atomic number 2, around which the two remaining electrons could revolve. One would thus have changed hydrogen into helium and in the process released energy. But fusing and transforming atomic nuclei is not that simple.

Arthur Eddington and the Energy Source of Stars

In 1926 Sir Arthur Eddington, Plumian Professor of Astronomy at Cambridge University, published his *Internal Constitution of the Stars,* a brilliant summation of the then available knowledge of the physics of stellar interiors, a field to which Eddington had made no small contribution. A theory explaining how stars functioned had been developed, but what was missing was the central clue, namely, the knowledge of how energy was produced in stars.

Stellar matter, so rich in hydrogen, was recognized as an ideal potential energy source.* It was known that in the transformation of

---

*Appendix A explains how we arrived at our knowledge of the chemical composition of the stars.

42

hydrogen into helium enough energy was released to cover the energy requirements of the sun and stars for billions of years. Obviously, a wonderful source of energy was available under the right conditions, but the conditions under which the fusion of hydrogen atoms takes place were not known. At the time, the experiments that would solve the mystery were still a thing of the future.

Astrophysicists had little choice but simply to accept as given that stars were gigantic nuclear power plants, for they knew of no other process able to provide enough energy to fuel solar radiation for billions of years. Eddington put it most logically. Alluding to the many and repeated measurements of stellar brightness conducted by astronomers, he said, "The measurement of liberation of sub-atomic energy is one of the commonest astronomical observations; and unless the arguments of this book are entirely fallacious we have a fair knowledge of the conditions of density and temperature of the matter which is liberating it."* However, physicists were convinced that the atomic nuclei of stars could not react with one another.

At that time Eddington was already able to estimate the temperature of the solar interior. The sun is held together by the gravitational force of its mass, which pulls matter toward the center. Solar matter is prevented from falling into the sun's center by the counterpressure exerted by solar gas, which pushes the matter outward and thus counteracts the gravitational force. The two forces are in a state of equilibrium, a phenomenon that also affects the atmosphere of the earth. Were it not for gravitational force, air pressure would blow our atmosphere into space; and were it not for air pressure, the air enveloping the earth would fall down onto its surface. In the case of the sun the gravitational force by which solar matter attracts itself can be calculated. The gas pressure must be as great as the gravitational force to be in equilibrium with it. The gas pressure depends on the density and temperature of the gas. We know the density of the solar mass because we know the sun's mass and volume. How great then is the pressure of solar matter? That depends on its tem-

*A.S. Eddington, *The Internal Constitution of the Stars* (Cambridge: Cambridge University Press, 1926), p. 296.

perature. The hotter a gas the greater its pressure. How hot must gas in the solar interior therefore be to maintain equilibrium with the sun's gravitational force?

Eddington estimated the temperature in the sun's central region to be about 40 million degrees. That may strike us as extremely high, but nuclear physicists thought it far too low to permit nuclear reaction. At that temperature the atoms in the solar interior move at the rate of about 1,000 kilometers per second. At that speed the hydrogen atoms have shed their electrons, and their protons fly freely through space. Occasionally, two will meet, but since both carry a positive charge they will repel each other. At the speed they are traveling they will come very close to each other but will be diverted by their repelling electrical forces before they can come close enough to fuse. Moreover, to make a helium nucleus out of the hydrogen atoms four protons and two electrons, a total of six particles, would have to meet simultaneously—an extremely unlikely event. Even if by chance all six were to move toward each other, electrical forces would divert them and prevent fusion. Only at temperatures of tens of billions of degrees would the particles approach each other with sufficient force to allow them to fuse despite the repelling electrical forces. Physicists were convinced that the sun, with its internal temperature of 40 million degrees, was too cold to let the hydrogen turn into helium. But Eddington was certain that nuclear energy was the only possible fuel of the stars. He wrote defiantly, "We do not argue with the critic who urges that the stars are not hot enough for this process; we tell him to go and find a *hotter place.*"* In his view the assumptions of the physicists about the conditions under which helium is formed from hydrogen were still too speculative. He put his trust in the stars and believed that ultimately the physicists would find out that hydrogen could change into helium at the comparatively low temperature of 40 million degrees. He turned out to be right.

*Eddington, *The Internal Constitution of the Stars,* p. 301.

George Gamow and the Tunnel Effect

At about the same time that Eddington was clinging to his notion of the stellar transformation of hydrogen into helium, a revolution in physics was being launched by Louis de Broglie in Paris, Niels Bohr in Copenhagen, Erwin Schrödinger in Zurich, and at Max Born's school of quantum mechanics at Göttingen. It was Göttingen's golden age. Among the many young physicists who came there from all corners of the globe were such future luminaries as Werner Heisenberg and Robert Oppenheimer, Paul Dirac and Edward Teller, and a young Russian named George Gamow who devoted himself to a particular aspect of radioactivity—the natural transmutation of atomic nuclei.

Certain chemical elements break down spontaneously: uranium turns into thorium, thorium into radium, and radium continues to undergo change. The nucleus of the most common type of radium consists of 88 protons and 138 neutrons. After a given period of time a radium nucleus discharges two neutrons and two protons and turns into a less massive atomic nucleus. The discharged particles remain bound together; they form a helium nucleus. How is it possible for a radium nucleus to give off a helium nucleus? The building blocks of the radium nucleus are packed into a very tiny space and held together by an extremely powerful force, *nuclear force*. This force is far more powerful than the electrical repulsion of the protons. Without it the radium nucleus, propelled by the mutually repelling protons, would fly apart. The range of nuclear force is, however, limited. If a part of the nucleus wanders too far afield, electrical repulsion predominates and the two parts split. According to traditional physics this cannot really happen since nuclear force holds the nucleus together. Yet it does happen in nature.

Gamow solved the problem of the transmutation of atoms. True, the components of a radium nucleus are bound together by nuclear force and cannot theoretically move apart, but quantum mechanics tells us that the probability of such a separation is limited. Contrary to traditional physics a part of the nucleus can, despite the powerful forces keeping it together, occasionally move far enough away from the rest of the nucleus for electrical repulsion to drive it still farther out. The process, unlikely though it is, nonetheless does occur. In the case of radium atoms it takes more than a thousand years until a helium nucleus is repulsed by a radium nucleus.

This phenomenon, known as the *tunnel effect,* was understood only through quantum mechanics. The label itself was the product of visual imagery. The building blocks of a radium nucleus are kept imprisoned by nuclear force as if by a ring of mountains that cuts them off from the surrounding world. They lack the energy needed to cross the mountains into freedom. According to traditional physics they cannot leap over the mountain, but quantum mechanics claims that a nuclear building block will occasionally get to the other side of the mountain by tunneling its way through. But if particles are able to penetrate the retaining wall from the inside then, so said Gamow, particles must also be able to penetrate the atomic nucleus from outside the wall.

The Tunnel Effect in Stars

Let us return to the stars and to the question of where their energy comes from. If radium nuclei do something they in fact ought not to be doing, why then shouldn't the protons in the sun also do something the physicists would deny them? In the case of radium, nuclear force is supposed to prevent the protons from moving so far apart that their electric repulsion becomes effective. But the radium nucleus breaks apart even though it really is not supposed to. Isn't it therefore possible for solar protons to fuse even though by right they ought not to?

The physicists Robert Atkinson and Fritz Houtermans solved the problem of stellar energy production with the help of Gamow's tunnel effect. In March 1929 they submitted a paper to the

Zeitschrift für Physik dealing with the structural possibilities of the elements in stars, which begins with these words: "Recently Gamow demonstrated that positively charged particles can penetrate the atomic nucleus even if traditional belief holds their energy to be inadequate."* The authors then went on to explain why hydrogen nuclei, which according to traditional physics can fuse only at temperatures of tens of billions degrees, could nonetheless come sufficiently close for fusion at the comparatively moderate temperatures of stellar interiors. Even though the protons of stars are separated from one another by electrical fields as if by the walls of mountains, they still manage—perhaps only after a very long time—to tunnel their way through despite insufficient energy. The likelihood is not very great, but the effect apparently occurs in the interior of both the sun and stars frequently enough to allow them to live off the energy thus released. Atkinson and Houtermans had proved what Eddington had only surmised: both sun and stars cover their energy needs via the transmutation of hydrogen into helium. Their paper laid the groundwork for the theory of thermonuclear reactions, the theory of the processes that ultimately are responsible for energy production in stars. The energy source of the sun and the stars had been found.

Robert Jungk, gathering background material for his book, *Brighter Than a Thousand Suns,* interviewed Houtermans, who reminisced about those days at Göttingen: "In the evening, after we had put the finishing touches on our essay," he told Jungk, "I went for a walk with a pretty girl, and as it grew dark star after star lit up the sky. 'How beautifully they twinkle,' exclaimed my companion. I puffed out my chest and said: 'And since yesterday I know why.' She didn't seem the least bit impressed. Did she believe me? Apparently at the time it was a matter of complete indifference to her." End of anecdote.

In 1965, when I joined the faculty at Göttingen, I thought I'd find out whether said lady still lived there. Like so many good intentions this one was never carried out. But I did meet her seven years later

*R. d'E. Atkinson and F.G. Houtermans, *Zeitschrift für Physik* 54 (1929): 656.

at a conference in Athens which the Atkinsons, who by now were living in Bloomington, Indiana, were also attending. Mrs. Atkinson told me that she had been the girl and that Houtermans had indeed mentioned his discovery to her but that the circumstances were not as romantic as one might think from Jungk's account. I learned something more important. I asked Doctor Atkinson what had set him off on that particular quest. He said that reading Eddington's book had brought to his attention the problem of the energy production of the stars, namely, that the temperatures in their interior were not high enough for nuclear fusion to take place. Yet Eddington was certain that nuclear energy and nothing but nuclear energy could fuel the radiated energy of the sun and stars. Atkinson discussed the dilemma with Houtermans. The timing was right: Gamow had just published his paper, the problem no longer appeared insoluble, and Atkinson and Houtermans set about finding the answer.

Since then we have learned that nuclear reactions can take place in stars. But what kind of nuclear reactions? Do protons fuse with protons, or do protons penetrate other atomic nuclei? And if one of these two, which one? The answer to these questions was not found for another ten years.

The Carbon Cycle

How does the hydrogen in the stars become transformed into helium? This question was first answered independently by Hans Bethe in the United States and Carl Friedrich von Weizsäcker in Germany. In 1938 both Bethe and von Weizsäcker discovered the first reaction that in fact turns hydrogen into helium and is able to cover the energy needs of the stars.*

*The time for this seemed to have been ripe: on July 11, 1938, von Weizsäcker's paper was submitted to the *Zeitschrift für Physik,* and on September 7, 1938, Bethe's manuscript reached the editorial office of *Physical Review.* Both papers contained the discovery of the carbon cycle. However, a paper by Bethe and Critchfield had been submitted to that journal on June 23, dealing with the most important part of the proton-proton chain. Of the latter more will be said in the next section.

The process is somewhat complicated; it presupposes that the stars, contain other elements in addition to hydrogen, for example carbon.

Carbon and nitrogen nuclei act as catalytic agents, familiar to us from chemistry. Hydrogen nuclei attach themselves to these catalytic nuclei, thereby in several steps forming a helium nucleus within the other nuclei that ultimately is released. The carbon and nitrogen nuclei themselves suffer no damage.

As we can see in figure 3.2, this is a circular process. Let us begin at the top of the diagram, where a carbon nucleus with the number 12—we refer to it as C^{12}—meets up with a hydrogen nucleus. The tunnel effect enables the hydrogen nucleus to overcome the repulsing electrical field of the carbon and to fuse with its nucleus. The new nucleus now consists of thirteen heavy particles, both protons and neutrons. The added positive charge of the proton has increased the charge of the original carbon nucleus; the nuclear-charge number becomes 7. We now have a nucleus of the element nitrogen with the mass number 13, and it is designated N^{13}. This type of nitrogen is radioactive and after a time gives off two light particles—a positron and a *neutrino.* * The nitrogen is thus transformed into C^{13}, carbon with the mass number 13. Its nucleus has the same charge as the original carbon atom, but its mass number is higher. We now have an isotope of the original nucleus. If another proton meets this new carbon isotope, we again get nitrogen, which, however, has the mass number 14; it is N^{14}. If still another proton meets this new nitrogen atom, it becomes transformed into O^{15}, oxygen with the mass number 15. This nucleus is again radioactive and gives off one positron and one neutrino, thereby changing into N^{15}, nitrogen with the mass number 15. Considering that the process began with carbon having the mass number 12 and has arrived at nitrogen with the mass number 15, it is obvious that the continuous addition of hydrogen nuclei has increased the weight of the atom. If still another proton is added to the nitrogen atom, then two hydrogen atoms and two neutrons

---

*We will discuss the strange role played by neutrinos in chapter 5.

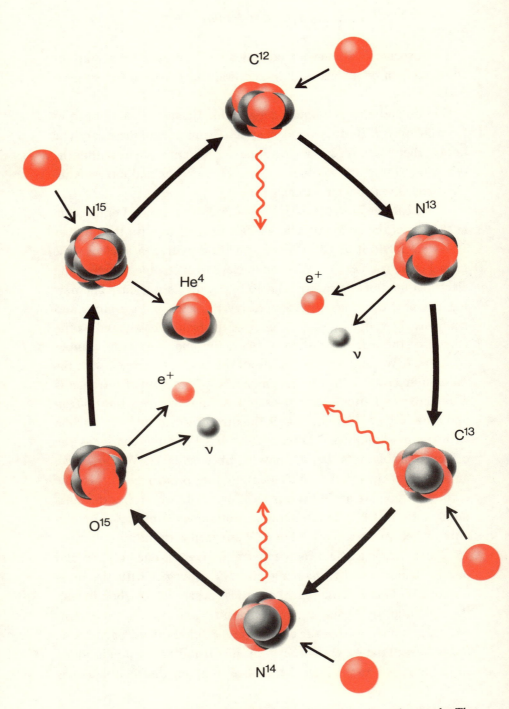

Figure 3.2 The transformation of hydrogen into helium in the carbon cycle. The color scheme is the same as in figure 3.1. The red, wavy arrows indicate atomic radiation; e^+ denotes positrons; ν, neutrinos.

are given off and together form a helium nucleus. With that our nucleus changes back into the original carbon nucleus, and the circle is closed.

In this process a total of four protons was swallowed up and one helium nucleus produced: hydrogen was transformed into helium. Via this process enough energy is released to enable the stars to shine for billions of years. The stellar matter is fueled by each step of this process. The radiation released by each reaction transfers energy to the stellar gas, and the union of the positrons with the free-flying electrons also creates radiation that helps fuel the stars. A small portion of the energy is carried off by the neutrinos.

In 1967 Hans Bethe received the Nobel Prize for Physics for his work on the genesis of stellar energy.

We know that the cycle requires the presence of catalytic elements—carbon and nitrogen. One of the isotopes appearing in the cycle suffices. The catalysts for the subsequent reactions are produced once a reaction is set off. Moreover, while the cycle runs its course the reactions even establish a definite quantitative ratio of these elemental isotopes. This relationship depends on the temperature at which the cycle operates. Today's spectroscopic devices allow astrophysicists to make fairly accurate quantitative measurements of cosmic matter (see appendix A). The quantitative ratio of the isotopes C^{12}, C^{13}, N^{14}, and N^{15} frequently allow the scientist to determine whether the matter they are measuring has been involved in the stellar fusion of hydrogen via the carbon cycle, and beyond that, at what temperature that fusion had taken place.

The carbon cycle is not the only process by which hydrogen can be transformed into helium. Another, simpler method is even more important—at least as far as the sun is concerned. This method was discovered at about the same time as the carbon cycle.

The Proton-Proton Chain

The carbon cycle requires a given amount of carbon or nitrogen. During this process the atoms of these elements are not depleted; they form a sort of temporary shelter in which hydrogen atoms are transformed into helium atoms over time. In 1938 Hans Bethe and Charles Critchfield demonstrated that this transformation can also take place in the absence of carbon and nitrogen.

The process is depicted in figure 3.3. Two protons collide and fuse. They discharge a positron and a neutrino. The residual nucleus then consists of only a proton and a neutron. It has the same charge as hydrogen but is twice as heavy; it is a form of heavy hydrogen called *deuterium.* When a hydrogen nucleus meets a deuterium nucleus they combine to form a helium atom consisting of two protons and one neutron. That helium is not yet the "proper" helium but the lighter isotope He^3; its nuclear charge is the same as that of helium but its mass number is smaller. When two such light helium nuclei meet they fuse into one "proper" helium nucleus, and in the process two hydrogen nuclei are freed. In this chain a total of four hydrogen nuclei have been integrated into one helium nucleus.

Which of the two processes, the proton-proton chain or the carbon cycle, takes place in the stars? The answer is both, if the temperature is high enough. At 10 million degrees the processes of the proton-proton chain are more prevalent. When the temperature rises significantly above that, energy production via the carbon cycle predominates.

The proton-proton chain may possibly have played a major part in the formation of the first stars. Today it is assumed that during the birth of the universe precipitated by the so-called big bang, only hydrogen and helium were formed, so that the first stars lacked the catalytic agents needed for the carbon cycle.* If that is so, hydrogen

*The history of the matter out of which the first generation of stars most probably was born is the subject of Steven Weinberg's *The First Three Minutes* (New York: Basic Books, 1976).

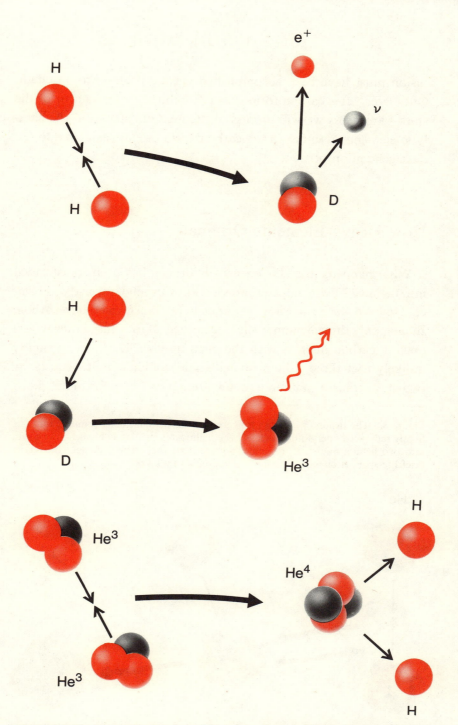

Figure 3.3 The proton-proton chain. The color scheme is identical to that of figure 3.2. Here, too, hydrogen is being transformed into helium. In the upper part of the illustration, two hydrogen nuclei collide and form a deuterium nucleus. In the center a deuterium nucleus and a hydrogen nucleus unite to form a helium isotope. At the bottom two nuclei of the helium isotope combine to form normal helium with the mass number 4.

fusion must have been accomplished via the proton-proton chain. Only later, after carbon formed out of helium in the interior of the stars—a process we will discuss in the next chapter—did the catalytic elements essential to the carbon cycle become available to the next generation of stars.

How Heavy Elements Originate

What happens in a star once its hydrogen has been transformed into helium? Edwin Salpeter, presently at Cornell University in Ithaca, New York, showed how helium can be transformed into carbon. In fact, only three helium nuclei are needed. If they fused they would make a carbon nucleus with the mass number 12. But it is highly unlikely that three helium nuclei would collide simultaneously. A two-stage transformation, as shown in figure 3.4, is far more likely:

Figure 3.4 The fusion of helium into carbon. Two helium nuclei combine into one highly radioactive beryllium nucleus, which almost immediately breaks down again into two helium nuclei. If it collides with still another helium nucleus during its brief life span, it changes into carbon and emits radiation.

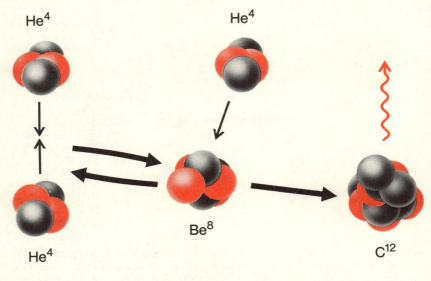

He^4 He^4

He^4 Be^8 C^{12}

two helium nuclei collide and form one beryllium atom with the mass number 8. This type of beryllium is highly radioactive; the life span of the newly formed beryllium nucleus is brief beyond belief. Within 10 millionths of a billionth of a second, it breaks down again into the two helium nuclei from which it originated. But if in the course of its brief life it collides with a third helium atom, good, stable carbon is produced. Almost without exception the beryllium nuclei break apart and only rarely is one of them saved from disintegration by a free-flying helium atom. But in stellar matter with temperatures of 100 million degrees such transformations are so numerous that the energy thereby released can fuel the star.

What happens afterward? How does the story continue? At somewhat higher temperatures carbon atoms fuse and in different ways break down into elements like magnesium, sodium, neon, and oxygen. Oxygen atoms can fuse and form sulphur and phosphorus, creating increasingly heavy atom nuclei. One might well ask whether in the final analysis all chemical elements were not perhaps brewed out of the hydrogen and helium in the centers of stars. We shall come back to this question in chapter 11. Suffice it to say here that nuclear reactions can take place in stars and that the transformation of hydrogen into helium under the conditions presumably obtaining in them can cover the demands of stellar radiation for long periods of time.

But what are the conditions in the interior of stars? How do we know their exact temperatures? We cannot look inside and do not have access to direct information. Yet we are fairly well informed about the state of a star's interior, better than about the interior of our earth. The reason for this and the important contribution computers have made to our knowledge of the stars will be discussed in the next chapter.

4

Stars and
Stellar Models

We are in the fortunate position of being able to look into stars so that we can learn something about their interior. Stars after all are not merely phenomena inviting our awestruck admiration. They are objects of our universe and as such are subject to the laws of physics, which was demonstrated by our applying to them the law of the conservation of energy and calculating how long they would be able to live off their nuclear energy. Not only the law governing energy conservation but all physical laws apply equally to the stars and every other part of the cosmos.

In this section I will outline how the laws of physics and the known properties of stellar matter determine a star's structure, so that with the help of a computer we can, so to speak, glimpse its interior. In the case of simple stars all we have to do is determine the amount and chemical composition of their matter; once that has been established, we can calculate a star's structure mathematically, via equations, without ever even looking at it in the sky. Not only can its surface temperature and luminosity be calculated and entered on the H-R diagram but also its diameter and even—and this is of particular interest—the pressure, temperature, and density of its interior. Readers who find this account too detailed may wish to skip the next few pages up to the section entitled "A Model of the Initial Sun." There we will start from the assumption that the physical laws and material properties of stellar matter that we are about to describe have been stored in a comprehensive computer program, and we will base our experiments on that program.

Gravity and Gas Pressure

Except for brief intervals stars must be in equilibrium. That equilibrium is maintained by the weight of stellar matter pressing down on the inner layers and the outward pressure of stellar gas. Were it not for gas pressure all stellar matter would collapse toward the center of the star, and were it not for gravity gas pressure would hurl all stellar matter into space. These two forces—gravity and pressure—must balance each other throughout the star, and their equilibrium allows us to calculate the gas pressure in the star. As we know Eddington used this balance to estimate the pressure in the center of the sun, and from that he was able to deduce its temperature to be 40 million degrees. If we want to apply this method to other stars, we must first know something about the gases of which they are made.

The material of which stars are made is not a mystery. They are composed of substances already familiar to us. We have tested and retested the properties of hydrogen and helium, the major components of stars, as well as those of the other chemical elements contained in them. Even though we are not able in our laboratories to duplicate at high temperatures the density of matter found in stellar interiors, still we know enough to estimate its properties. In this context a fortuitous circumstance proved of help. We on the earth are used to low-density gases. If the air of the earth's atmosphere or some other gas is compressed to the density of water or to an even greater density, its pressure changes in a complicated fashion. The gas can liquefy or solidify, at which point its properties become still more complicated. For in the event of great compression, the atoms come

57

very close to one another and their electron shells act as mutual obstructions. Exactly how this mutual blockage of different atomic shells comes about is still not fully understood. But as a result we lack exact information about the properties of matter in the center of the earth and know so little about its interior.

But in the stars things are different. In them high temperatures are the norm. Where matter is densely packed the temperature is so high that atoms long ago lost their electron shells. The connection between electrons and atomic nuclei has been broken; nuclei and electrons move about freely. These particles now require far less room than did the electrically neutral hydrogen atoms composed of both atomic nuclei and electrons. That is why hot stellar matter behaves like a rarified gas despite its high density, in which a cubic centimeter contains 100 or more grams of matter. And that is why we know more about the sun's interior than about the earth's. Even when stars increase in density—again because of their high temperatures—the properties of their gaseous matter remain known to us. Only when stellar matter cools and atoms begin to crystallize do its properties become more complex. But that becomes a crucial factor only in some stars, primarily in the low-temperature white dwarfs.

Energy Production and Energy Transfer

Temperatures in the stellar interior are high enough for nuclear reactions to take place and for the subsequent release of nuclear energy. In the 1920s and 1930s Atkinson and Houtermans, Bethe and von Weizsäcker showed us how atomic nuclei in stars react with one another; since then, other nuclear physicists have given us the information that allows us to calculate how much energy is released through nuclear reaction by a gram of stellar matter of a given density and temperature.

The hot core of the star emits energy that must escape to the outside. It must penetrate the outer layer of the star, generally in the form of radiation. Thus, opacity with regard to light and heat radiation is an important attribute of stellar matter. Principally in the outer layers of the stars, where the atoms have not yet shed their electron shells completely, photons radiating outward from the stellar core are absorbed by the remaining electron shells. That is to say, they are swallowed and after a while are again discharged. The outward-radiating photons jump from atom to atom, are absorbed, emitted, and then deflected and reabsorbed, and only after overcoming numerous obstacles and false starts do they finally get to the surface, from which they can fly away easily. The opacity of stellar matter therefore plays a most important part in the structure of a star, and its determination involves a series of complicated calculations. Astrophysicists are in the enviable position of having atomic physicists do the job for them, for the absorptive properties of atoms are vital factors in the work of atomic scientists.

After World War II unexpected help came to astrophysicists from another quarter. When an atomic bomb explodes, intensive light and heat radiation develop at the center of the explosion. This radiation is absorbed and reemitted by the atoms of the neighboring air masses. Therefore, atomic scientists, to predict the likely effect of an exploding bomb, must know the exact permeability properties of gases in relation to light and heat radiation.

Despite the secrecy surrounding these calculations, partial publication of the results is permitted and thus the findings are made available to astrophysicists. There is a team of scientists at the Los Alamos National Laboratory working in pure astrophysics, and their astronomical colleagues in both the East and West have access to their work on the opacity of stellar matter at specific densities and temperatures. As a further indication of successful East-West cooperation in astrophysics, the Los Alamos team has published some of its findings in a journal of the Soviet Academy of Sciences.

Seething Stellar Matter

The power of the outward radiation of a star is at times so great, and the opacity of the stellar matter so high, that energy builds up in the star's interior. When that happens the star finds another way to release the pent-up energy. The process is similar to one familiar to us from everyday life. We know that when a stove top becomes hot it radiates some of its energy into the room, but another type of energy transfer also takes place. The air above the stove heats up, expands (that is to say it diminishes in density), and rises to make room for cooler air masses. This warm air transports energy from the stove to other areas in the room. We refer to this type of energy transfer as *convection*. From open fires, from asphalt surfaces heated by solar radiation, heated air masses rise up and transfer heat upward; cooler air masses descend from above to warm themselves, and after a while they too rise up. Because convection plays a major role in the energy budget of the earth's atmosphere, meteorologists began to study it even before astrophysicists did.

In many stars matter begins to seethe when radiation fails to do its job, and convection must take over the transfer of energy. In the external layers of the sun, energy is carried outward not via radiation but by the motion of heated gas blobs. The bubbling gas of the sun can be seen even with small telescopes equipped with powerful filters to cut the glare. The sun's surface is not uniformly bright; hot, bright gas masses about 1,000 kilometers in diameter can be seen rising next to descending cool, dark masses. Figure 4.1 shows a photograph of the solar surface with the typical changing, blotchy structure that astronomers call *granulation*. Thus, the phenomenon of convection familiar to us from life on Earth plays an important role in the world of stars as well.

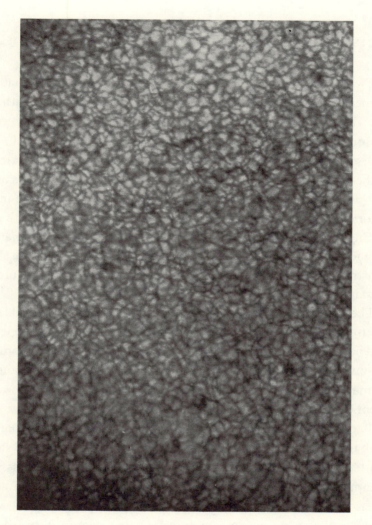

Figure 4.1 The granulation of the solar surface. Since the internal energy of the sun is transported to its outer layers via convection, hot and hence more luminous gas masses rise from the sun's surface, whereas cooler (and as shown here, darker) gas masses sink. These movements produce a constantly changing granulation on the sun's surface. In this picture 14 millimeters is equal to the earth's diameter. (Photograph: D. Soltau, using the 40-centimeter vacuum reflector of the Kiepenheuer Institute for Solar Physics, Freiburg, in Izana, Tenerife)

Stars in the Computer

I have mentioned only a few examples of the laws and material properties that help us understand the interior of a star. With our knowledge, most of which predates World War II, we can try to figure out the structure of a star without ever leaving our desk. The first scientist to have done just that was Robert Emden, a professor of thermodynamics at the Munich Institute of Technology. His book, *Gaskugeln* (Gaseous Spheres), published in 1907, is a classic in the theory of stellar structure. He was followed by Arthur Eddington in England and still later by Thomas Cowling and Subrahmanyan Chandrasekhar. In the 1920s and 1930s Cowling and Chandrasekhar constructed stellar models that gave a rough picture of the interior of stars.

The development of modern computers has made it possible for us to reopen this line of investigation by simulating the properties and behavior of stars electronically. What does that mean? Computers must be taught the laws governing the structure of stars, and information about the material properties of stars must be stored in them. In other words they must be fed data about the pressure of stellar gas at various densities and temperatures, made acquainted with the laws governing the change of hydrogen into helium in stellar matter, and told how much energy is released in the process. The computer must further be told how the energy released in the stellar interior penetrates stellar matter and reaches the surface and at what point radiation or convection transfers that energy outward. All these bits of information must be combined into one big computer program. We are then able to construct stars in computers and theoretically to follow their evolution. The computer printout gives us the temperature, density, gas pressure, and energy emitted by the various layers of a star. It describes the structure of a star at a given moment in time. We then say that the computer has given us a *stellar model.*

A Model of the Initial Sun

Let us assume that we have a computer program like the one described above and a computer big enough to handle it. Let us now embark on constructing stellar models. To begin with we have to decide on the chemical composition of our stellar matter. Suppose we decide on the mix of chemical elements found in the sun and in almost every other star; one kilogram of stellar matter would therefore contain 700 grams of hydrogen and 270 grams of helium, and the remaining heavier elements, above all carbon and nitrogen, would account for the remaining 30 grams. The computer is asked to determine the material properties, above all the radiative opacity of the stellar matter, for precisely this chemical mix. All it then needs to know is the amount of matter we want to assign to our stellar model. Let's assume that we decide on about the mass of the sun. With the help of the laws of nature stored in its memory and the known properties of stellar matter, the computer constructs a model. Modern computers are so rapid that they can come up with the solution to this problem in less than a minute. How does this solar model look? The computer model compared to the sun's data is somewhat smaller than our sun; its diameter is only 92 percent of the sun's, and it radiates less energy than we had expected. Its luminosity is only about 75 percent of the sun's, and its surface temperature of 5,620° is about 180° lower. For the time being we will ignore these differences and take a closer look at the solar model. In the H-R diagram it lies well along the main sequence, somewhat below the actual sun.

Figure 4.2(a) shows the inside of our solar model.* In the course of this work we will come across this type of representation again and again. For a detailed explanation of the model and its representation see the caption.

*Although many astrophysicists before and since Kurt von Sengbusch have constructed solar models, I am basing my figures on those in his dissertation submitted at Göttingen in 1967. The subsequent passages describing the evolution of the sun are also based on his findings.

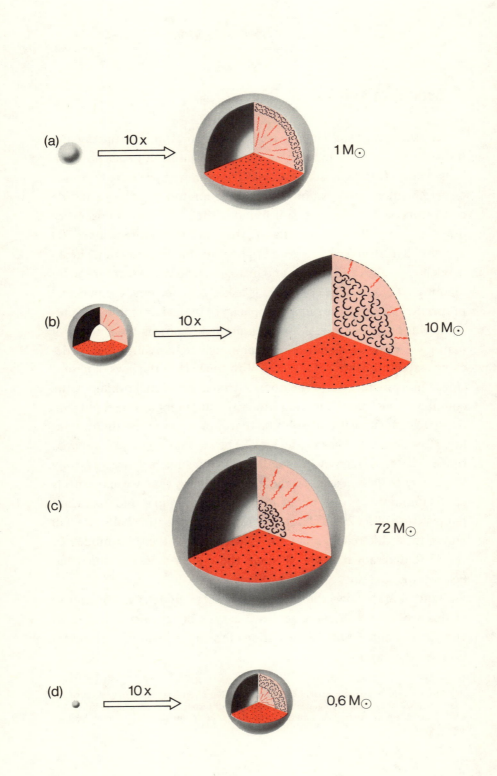

(a)

10 x

1 M⊙

(b)

10 x

10 M⊙

(c)

72 M⊙

(d)

10 x

0,6 M⊙

The density of matter in the center of our model amounts to 100 grams per cubic centimeter, or about 13 times that of solid iron. Its pressure equals 130 billion atmospheres, and the temperature in the central region hovers at around 10 million degrees. At that temperature nuclear reactions take place: nuclear energy is produced via the proton-proton chain. We now have a star that covers its luminosity through the fusion of hydrogen. Energy is transported from the interior to the outside by radiation, but this transfer mechanism does not operate in the outer layers. There, energy is brought to the surface by convection. Gas masses rise and fall, as in the illustration of the granulation on the surface of the sun.

To sum up: Out of matter of the same chemical composition and mass as the sun, we have built a star that is located on the main sequence in the H-R diagram, that turns hydrogen into helium in its interior and that in its outer layers is, like the sun, convective. In general it resembles the sun.

Yet given all this why didn't we get an exact replica of the sun? What accounts for the disparities? Was there something wrong with our computer program? As we shall see our model differs from the sun because our solar matter was of a uniform chemical composition. We failed to take into account that the actual sun has after all been shining for more than 3 billion years and that newly formed helium must have been building up in its central region for quite some time. We constructed a sun and proceeded as though the chemical compo-

---

Figure 4.2 The internal structure of stellar models of different mass. $M_\odot$ is the symbol for solar-mass units. In the models (a), (b), and (d), the stars at the left are drawn according to the same scale, as in (c). To show the internal structure more clearly, the models in (a) and (d) have been enlarged tenfold; in (b) only the inner portion has been enlarged. The three cross-sections represent chemical composition (bottom), energy production (upper left), and energy transport (upper right). The dots in the lower of the three cross-sections indicate the regions in which the chemical elements are still in their initial abundance. Thus, in all the models the original hydrogen-rich mixture still dominates. Light areas in the upper left portions designate sites in which energy is being released through nuclear reactions. Wavy arrows in the upper right portions indicate areas in which energy is being transported through radiation; and cloudy portions, areas in which it is being transported to the outside through convection.

sition of its center and its outer layer were identical. The sun we built had only just begun its nuclear burning; it was a sun in its infancy. What we came up with was the *initial sun*. Before we attempt to find out how we get from the initial sun to our sun, let us make the same computer calculation for stars of different mass but of identical chemical composition.

Discovering the Initial Main Sequence

Suppose we ask the computer to construct a stellar model of the same chemical composition and twice the mass of the sun. Within less than a minute it will come up with a printout of such a model, representing a star that also lives off hydrogen fusion. Whatever the mass of our programmed model, we find that the resultant stars all live from hydrogen fusion. Yet while stars of about 1 solar mass and less receive their nuclear energy via the proton-proton chain, in more massive ones the carbon cycle is responsible for converting their hydrogen into helium.

The computer also tells us the luminosity and surface temperature of the various models, which enables us to enter them on the H-R diagram (see figure 4.3). What do we find? All of them lie along a descending line leading from the upper left to the lower right: the most massive are on top, the least massive below. What we did was rediscover the main sequence, not through the observation of stars but through computer printouts of hydrogen-consuming stars of different mass. The computer substantiated our assumption, based on their life span, that the sun and other main-sequence stars cover their luminosity through hydrogen fusion. The main sequence is that section of the H-R diagram populated by the stars that derive their nuclear energy from hydrogen.

Our stellar models reflect still another property of the main-sequence stars alluded to earlier: the observed connection between mass and luminosity. The luminosity of a stellar model of 10

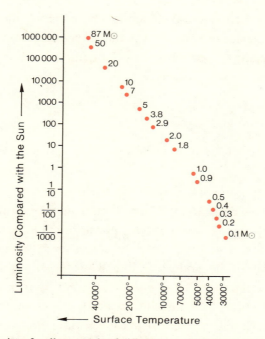

Figure 4.3 A series of stellar models of different mass (all composed of the identical hydrogen-rich mixture) form a main sequence in the H-R diagram possessing all the attributes of the observed main sequence. The mass of the individual models is given in solar-mass units ($M_\odot$). The radiation of the main-sequence stars increases substantially with greater mass.

solar masses is substantially greater than that of a 1-solar-mass model, and this increase in luminosity conforms to the observed mass-luminosity relationship as seen in figure 2.4.

What holds true for the sun also holds true for all our computer stellar models. They represent stars at the early stages of hydrogen fusion. They are thus *initial stars.* Consequently, their main sequence is not that of the stars we see in the sky but the main sequence of initial stars—it is the *initial main sequence.* However, because stars—so long as they do not manifest pronounced signs of depletion—do not change too radically, there is no significant difference between the initial main sequence and the observed main sequence.

Since the observed phenomena of stellar models conform to actual stars, there is reason to believe that these models also give us a reliable picture of their interiors. The models therefore offer us a view of the stellar interior not available to the observing astronomer. Having taken a look at the sun, let us now look into two other stars—one massive and one not so massive.

The Interior of Spica

To exemplify a massive star let us construct a model of 10 solar masses. Since Spica is about 10 solar masses, the computer model of this mass should replicate Spica's properties, and in fact, its surface temperature and luminosity do correspond to Spica's. What does the interior of the model look like? The temperature at its center is 28 million degrees, and the energy released by it is the product of the carbon cycle. Its luminosity is produced in an area one-fifth the diameter of the star itself. It produces more energy than can be disseminated through radiation. At this point convection comes into play. The innermost 22 percent of the star's mass is in a convective state [see figure 4.2(b)]. For the rest, energy is disseminated via radiation. Photons push outward, repeatedly are stopped in their track and diverted by atoms and electrons, until finally they reach the surface to give the star its luminosity. The density at the star's center amounts to a little less than 8 grams per cubic centimeter. Thus, the density of the still gaseous stellar matter equals that of solid iron. The pressure produced by the weight of the stellar matter on the star's center amounts to 35 billion atmospheres. That is the situation in the interior of Spica, the brightest star in Virgo.

All stars which like Spica are substantially more massive than the sun transport the energy in their central regions via convection, as shown in the 72-solar-mass model of figure 4.2(c). It should be noted that the more massive main-sequence stars also have bigger diameters.

We have already discussed stellar models of the initial sun and of a star more massive than the sun. Let us now turn to a star that is far less massive.

The Red Dwarf in Cygnus

The star 61 Cygni, located in the constellation Cygnus, owes its fame to Friedrich Wilhelm Bessel who, sometime between 1837 and 1838, used it as a vehicle in his experiment with a new method for determining distance (see appendix B). It is in fact a binary system: two stars, one of 0.5 and the other of 0.6 solar mass, revolving once every 720 years around their joint center of gravity. The star that interests us here is 61 Cygni A, the more massive of the two. It is a main-sequence star with a surface temperature of 3,700°. Smaller and considerably cooler than the sun, it is classified as a red star: it is a *red dwarf.*

If one constructs a stellar model of 0.6 solar mass, it has approximately the same external characteristics as 61 Cygni A and is located at approximately the same place on the H-R diagram. What does the interior of the red dwarf look like? Figure 4.2(d) shows us such a model. The temperature in the center amounts to a mere 8 million degrees. All its nuclear reactions participate in the proton-proton chain. Its core density of 65 grams per cubic centimeter is lower than the density at the center of the sun. The central pressure of 75 billion atmospheres is similar to Spica's. Energy in the interior is transported via radiation. As in the sun convection takes place in the outer layers, but in a far thicker area. Red stars typically have thick external convection zones.

The farther down one travels along the main sequence, toward cooler and hence redder dwarfs, the thicker the outer convection zone. All the matter of a star of only some tenths of a solar mass, from the outer surface to its center, is in convective motion.

Characteristics of the Initial Main Sequence

Now that we have gained a rough understanding of the properties of main-sequence stars, we have made a giant step forward, for more than 90 percent of all stars fall into that category. We now know, for example, that all of them owe their existence to the conversion of hydrogen into helium. The properties of the hydrogen atom determine the star's energy budget and consequently the external features of main-sequence stars. Since color and brightness are among these features—that is, characteristics visible in the sky—we can justifiably say that the stars we see project the properties of the hydrogen atom onto the sky, and if that atom were constructed differently, the stars would look different.

How far downward, to what low level of mass, does the main sequence extend? Can nature build a star that lives off the fusion of hydrogen out of any random quantity of concentrated hydrogenous matter? Can there be stars of a mass no greater than the mass of a human being?

If one instructs the computer to build a series of models of decreasing mass, starting with 1 solar mass, the central temperatures in these models would also decrease progressively. After a while the proton-proton chain would no longer function. The fusion of two He^3 nuclei (see figure 3.3), and therefore the change of hydrogen into He^4, would no longer be possible. At something like 8/100 solar mass, stars no longer burn hydrogen. Their internal temperature would not be adequate for the fusion of hydrogen nuclei. Main-sequence stars, which owe their existence to hydrogen fusion, must therefore be composed of at least 1/10th of a solar mass. At that point the main sequence comes to an end. The computer, if asked to construct models of hydrogen-burning stars of lesser mass, will simply balk. Suppose we were able to launch a vast space experiment to create a star composed of 1/1,000th of a solar mass; we might then come up with any type of object, possibly a planetary body, but certainly not a hydrogen-burning ministar.

What do things look like at the massive terminus of the main sequence? What happens if we were to ask the computer to build stars of a hundred or a thousand or even a billion solar masses? It would undoubtedly come up with such huge models, but they would all share a strange characteristic: if for brief periods they were to be lightly squeezed together, their centers would become significantly

denser and their temperatures would rise. This increase in temperature would stimulate hydrogen fusion, which in stars like these follows the carbon cycle, so that the energy released in the process
would push the compressed stellar matter toward the outside. As
a result the central region would cool off, nuclear energy production
would diminish, gas pressure would go down, and gravity would retrieve the outward-moving matter. This returning mass would in
turn compress the central region, and the game would start all over
again.

More detailed calculations about this process have been carried
out by a number of scientists, among them Immo Appenzeller at Heidelberg, and they show that these fluctuations would continue to accelerate until at each expansion a small portion of the external stellar
layer would be catapulted into space beyond retrieval at enormous
velocities. The star would lose mass with each fluctuation until our
superstar was reduced to about 90 solar masses, at which point this
circuitous process would come to an end. The central region would
cease to heat up so greatly under compression, the nuclear processes
would not respond with such enormous overproduction, and the
fluctuations would cease. The star would turn into an ordinary
main-sequence star of 90 solar masses, serenely burning up its
hydrogen.

One might of course interject that all this can take place only if,
as initially assumed, someone squeezes our superstar, for only then,
according to our scenario, does the accelerating cycle of expansion
and contraction set it. Fortunately, no one in the universe goes
around squeezing the stars. However, the fact remains that even the
slightest deviation from the equilibrium of gravitational force and
gas pressure can set off such fluctuations. But our universe is full
of disturbances. Even in the absence of any outside interference, the
movement of atoms in a star's interior or the movement of stellar
matter in areas where energy is transported outward by convection
could set off fluctuations until the star has lost enough matter to stop
the process.

This phenomenon sets a natural upper limit to the main sequence.

That, too, is in line with our observations. No stable star of a substantially greater mass than the theoretically predicted upper limit of 90 solar masses has ever been discovered.

Although our computer came up with pretty good simulations, the models were merely of initial stars just being formed. The hydrogen supply in their central region will soon begin to diminish, first in the more massive stars but subsequently in those of lesser mass as well. The stars will begin to age. In the next chapter we will investigate this process by looking at the model of our sun.

5

The Life Story
of the Sun

Helium is the residue of hydrogen's burning. While the surface of the initial sun is radiating energy into the cosmos, hydrogen is being converted into helium in its interior. With the passage of time hydrogen continues to be used up. In constructing the model of the initial sun, we assumed that is is uniformly composed of hydrogen-rich matter. But the build-up of newly formed helium in the solar center plays havoc with our simple computer models

From the Initial Sun to Our Sun

When we build a model of a main-sequence star, we learn how much energy is produced throughout its interior by the fusion of hydrogen. Thus, we also know the production per second of helium. At the center of the initial sun, one ten-millionth gram of helium is formed per kilogram of matter per year. By figuring out the amount of helium produced at each point in the star in a million years, we arrive at the chemical composition of a solar model that describes our sun a million years after the onset of hydrogen fusion.

Because of the changed chemical composition of the central region, the computer must now be asked to calculate a revised model, for those parts of the star in which the amount of helium has increased its material properties—that is, its opacity and the amount of hydrogen available for nuclear reactions—have also undergone a change.

This revised model describes how the sun looked a million years after the onset of nuclear reaction. It differs only slightly from the initial sun, for a million years are as nothing compared with the billions of years that the sun has been living off its fuel supply. Hence, the surface temperature of the new model is still almost the same as that of the initial sun, but its luminosity is somewhat greater. Even though less hydrogen is now available at the center, the center temperature of the new model has risen slightly, with the result that now even more energy is being produced there than previously.

The new solar model also tells us where energy is being released and how much hydrogen is being converted per second. That information enables us to predict this sun's chemical composition a million years hence, and so we ask the computer to construct a new model for this new mix of chemical elements.

We can now set up a series of models, and since we know both the surface temperature and luminosity of each, we can record them on the H-R diagram. Starting from the initial sun in that diagram, we can plot a chain of dots that depicts the evolutionary path of the sun through the diagram. This path is shown in figure 5.1, as are specific moments in elapsed time since the onset of hydrogen fusion.

The evolutionary path of our computer sun intersects the place in the diagram at which our sun is located today, which demonstrates that the difference between the properties of the initial sun and our sun is a consequence of evolution. Only after helium has become enriched at the solar center does the model acquire the properties of our sun. This tends to support our faith in the accuracy of our calculations and furnishes us with a clue to the age of our sun. The chain of models from the initial sun to our sun spans 4.5 billion years, the age of our sun. That is how long it took for the initial sun to turn into our sun. Before we go on to see what the future holds, let us spend a little more time with the sun as we know it today.

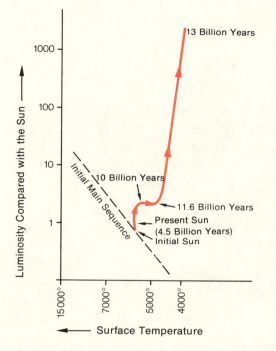

Figure 5.1 The evolution of the sun in the H-R diagram. From the initial sun the evolutionary track leads via the present sun from the initial main sequence to the area of the red giants. The years indicate the time elapsed since the ignition of hydrogen in the initial sun.

The computer gives us a chance to look into the interior of the sun. Figure 5.2(b) shows a model of our sun today. When we compare it with that of the initial sun in figure 4.2(a), we find that both have convective outer layers and both transport energy from the interior to the outside via radiation. In both, hydrogen fusion is accomplished by the proton-proton chain. The central region of the initial sun differs from that of our sun, where helium is found in larger concentrations. Whereas a kilogram of matter in the outer layers of the sun contains only 270 grams of helium, its center contains 590 grams per kilogram of matter; in other words about 300 grams of helium per kilogram of solar matter have been formed since the onset of hydrogen fusion.

In the outer layers solar matter is in a continuous state of motion.

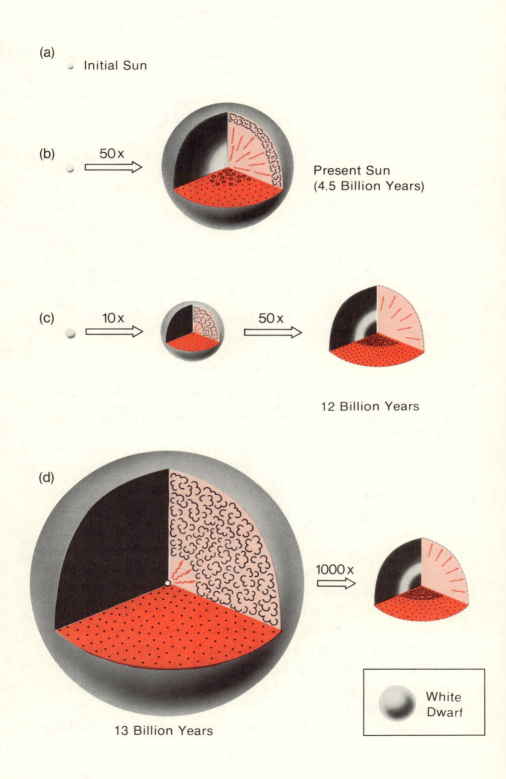

(a) Initial Sun

(b) 50x → Present Sun (4.5 Billion Years)

(c) 10x → 50x → 12 Billion Years

(d) 1000x → 13 Billion Years

White Dwarf

Every gram of matter that floats to the surface at one time had rested on the bottom of this seething layer, which at a temperature of 1 million degrees is 170 times as hot as the surface. Confirmation that the convective zone at the surface actually extends down into these hot areas has come to us from an entirely different source.

Where Is the Sun's Deuterium?

The atomic nucleus of deuterium, a hydrogen isotope, contains one proton and one neutron, and it is not very heat-resistant; a temperature of $500,000°$ is enough to fuse it with an ordinary hydrogen nucleus into a helium isotope. In nature deuterium appears only in small quantities, as for example in interstellar matter, the breeding ground of all stars. Deuterium must have been present when the sun was born, considering that traces of it are found on Earth. The waters of the oceans normally contain one deuterium atom for each five thousand atoms of ordinary hydrogen.

But deuterium is not found in the sun's atmosphere. Its absence is not surprising, since we know from our computer model that there is convection in the outer layers of the sun. Deuterium atoms on the solar surface would sooner or later be carried to the bottom of the

Figure 5.2 The internal structure of solar models in various stages of evolution. The representational scheme is the same as that in figure 4.2. Because in this illustration we are dealing with areas in which helium has become enriched, small circles are here used to indicate newly formed helium. Some are still mixed with the initial hydrogen-rich matter, which is indicated by dots. Later, only helium is found in the central region. The models on the left are all drawn to the same scale (but not the scale used in figure 4.2); in those at the right the inner part is enlarged; the enlargement factor is given. Model (a) is the initial sun, (b) the present sun. The helium core in model (c) has formed after the depletion of the hydrogen. The nuclear burning now takes place in a thin layer around the helium sphere. Model (d) shows the sun as a red giant with a thick outer convective zone and a comparatively small helium core whose dimension is reminiscent of a white dwarf. For purposes of comparison a white dwarf is shown to the right. It is drawn in the same scale as the interior of model (d), which is enlarged 1,000 times relative to the picture of model (d) on the lower left.

convective zone, where temperatures of 1 million degrees prevail, via the rise and fall of matter. Long before ever getting to the bottom they would change into helium with the help of hydrogen nuclei. Therefore, all deuterium in the sun must have been destroyed a long time ago. Even if deuterium were to get to the sun from somewhere in space today, it would be carried downward by convection and destroyed within two or three years.

The Lithium Problem

Our computer models cannot furnish us with an explanation for everything. In studying the chemical composition of the solar surface, we find that the sun, unlike the earth, is sadly deficient in still another element, *lithium*. Lithium, one of the lighter elements—three protons and four neutrons generally make up its atomic nucleus—is a rare commodity on the sun. A kilogram of solar matter contains one-hundredth the amount per kilogram found on Earth or in the matter falling to Earth in the form of meteorites. Is it possible that this element too was destroyed by the high temperatures on the bottom of the convective layer?

Lithium can in fact absorb a hydrogen nucleus and turn into two helium atoms, as shown in figure 5.3. But the million-degree temperatures of the layers into which the lithium atoms of the solar surface would descend are not high enough to facilitate this conversion; the destruction of lithium would require 3 times that amount of heat. Since all computer models, from the initial sun to our sun, fail to show us any convective zones deep enough to reach such temperatures, our calculations cannot explain the dearth of lithium on the sun. Was there perhaps a shortage of lithium from the very outset? That seems highly unlikely. It is generally assumed that the sun, the planets, and meteorites were born of the same type of matter, and thus were originally of the same chemical composition. We will come

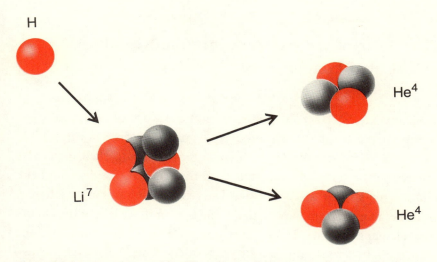

Figure 5.3 At temperatures of 3 million degrees lithium atoms in the stellar interior are transformed into helium with the help of hydrogen nuclei.

back to this point when we take up the origin of the stars. But if our assumption is correct, what happened to the sun's lithium? Where do we turn for the answer?

The key to the solution goes back to the birth of the initial sun, predating its burning of hydrogen. At that time the solar convective zone reached down far more deeply into the hot solar interior, entering regions with temperatures of at least 3 million degrees. At that time a large portion of the lithium in the sun's outer layers commingled with the interior matter and was destroyed. We will touch on this again in chapter 12. But before doing so we have to find out what there was before there was an initial sun. For the time being we are still trying to study the aging process of the sun. We will leave its youth for a later discussion.

It is only recently that we have learned more about the fate of sunlike stars of about the mass of the sun after they deplete their hydrogen supply, as shown in the evolutionary graph of figure 5.1. Not until the 1950s were computers used on a large scale to track the evolution of stars. Before going into the results of these calculations, I would like to add some historical and personal notes.

The Year of the Advance into the Realm of the Red Giants

The year 1955 saw the publication of the work of two great contemporary astrophysicists, Fred Hoyle and Martin Schwarzschild. Too bulky for inclusion in the *Astrophysical Journal,* their paper appeared in the series of supplements to this journal. Hoyle, who held Eddington's chair at Cambridge University, had already written a number of important papers, including one on the origin of the chemical elements in stars. In his spare time he also wrote science fiction. Martin Schwarzschild was four years of age when his father, the astronomer Karl Schwarzschild, died. His interest in astronomy dates back to his childhood. Apparently, the only thing that posed a threat to his scientific career was his childhood dream of becoming a milkman. But ultimately he turned to astronomy, largely because, so he says, he lacked the originality not to follow in his father's footsteps. He graduated from the University of Göttingen in 1935. The Schwarzschilds and the Rothschilds were allegedly neighbors in the old Frankfurt ghetto. The young Jewish astronomer thought it advisable to leave the Third Reich at the earliest possible opportunity; his brother, who remained in Germany, later committed suicide. Martin went first to Norway and from there to the United States. After the war he was appointed to a professorship at Princeton University.

Schwarzschild's Princeton group built main-sequence models to study the behavior of stars once they had used up the hydrogen at their centers. The great breakthrough came in 1955, with a work that for the first time showed how stars, starting out from the main sequence, turn into red giants.

Computers were then beginning to be used more and more frequently in astrophysical research, and Hoyle and Schwarzschild used them in their simulation of the evolution of stars. Soon afterward I was given the opportunity to follow in their footsteps in this field.

In the fall of 1957 Stefan Temesvary (1915–1984), then at Göttingen, spent many a night in front of the G-2, a computer built by Heinz Billing and his colleagues at the Max Planck Institute for Physics. Factory-built computers were not so readily available as today, and so the various scientific institutes built their own. Today we have portable electronic computers that can do the same things as our old model, which took up an entire room and whose tubes moreover heated up the room. Ludwig Biermann, then the head of the astrophysics department at the institute, suggested that we repeat the Hoyle-Schwarzschild calculations on our computer, using a revised method of our own.

A comparison of our old methods and today's approaches dramatically illustrates the rapid strides that have been made in the interim. To construct stellar models we had to begin with some arbitrary values for luminosity and surface temperature and step by step work from the surface downward until we arrived at the center of the star, only to find out that the model at that point had become meaningless, or, expressed in the language of mathematics, that the center could not fulfill the internal boundary conditions. We then had to recalculate the entire problem, using revised values for luminosity and temperature, in the hope that the internal conditions would now be approximated more closely. Thus, it took many "integrations" from the surface to the center of the star before we arrived at a sensible model. Our work took us on many journeys through the star, each a five-hour trek, and we prayed that the computer would continue to work for five hours without breaking down, else we would have to begin all over again. Today the computer of the institute, which has since moved to Munich, can construct a stellar model in seconds, a feat made possible not only by a far more sophisticated computer technology but also by the work of one man and his colleagues. But more of that in the next chapter. For the time being let us confine ourselves to examining the depletion of the hydrogen supply in sun-like stars. It is the fate of our sun and, as we shall see, it also bears on our future on this planet.

The Future of the Sun

What does the future hold? What will happen as more and more hydrogen is converted and more and more helium continues to form at the center of the sun? Computer calculations show that in the next 5 billion years nothing much will change. As we can see in figure 5.1, the sun in the course of its evolution slowly moves up the H-R diagram; that is to say, it approaches greater luminosities, becomes a little hotter on its surface, and subsequently cools off somewhat. Still, not many changes are to be expected.

Ten billion years after the onset of hydrogen fusion in the initial sun the sun's luminosity will be approximately twice what it is today. Mankind, if it still exists by then, will have endured severe climatic crises, but worse will be in store. To begin with, the sun will have grown to twice its present size. Meanwhile, major changes will have taken place in the sun's interior. The hydrogen at its center has been used up; its center region is taken up by a core of helium (see figure 5.2(c) for 12-billion-year model). Nuclear burning cannot take place in it since all the hydrogen has been used up and the temperature is too low for helium fusion (see figure 3.4). The only place where hydrogen fusion still takes place is on the surface of the helium core, where helium and hydrogenous matter meet. As hydrogen is burned, the newly formed helium is incorporated into the growing helium mass. Whereas up to then our sun had had a hydrogen-burning central region, it now has a core whose hydrogen-burning shell continues to eat into the still hydrogen-rich outer layers. With the passage of time the helium core at the center continues to increase in mass.

As we can see in figure 5.1, the star now moves toward the upper right of the H-R diagram. The sun ball continues to grow and at the same time cools off slightly. After 13 billion years the sun will have grown to about a hundred times its present size, and its luminosity will be about 2,000 times as great [see figure 5.2(d)]. At the same time its surface temperature will have dropped to 4,000°, 1,800° below its present level.

The Life Story of the Sun

But mankind is beyond salvation. The oceans of the earth have long since evaporated; lead melts in the intensity of the sun's light. The earth has become a cauldron unable to sustain life. A gigantic red sun spreading across more than half the sky shines down on an earth devoid of all life. Now we must ask whether the computer's projections are in fact an accurate picture of the future.

Our observations have properly described the essential properties of our sun. Does that mean that we are justified in assuming that these alarming predictions will indeed be fulfilled? There is direct evidence that they will. Looking at the H-R diagram of a globular cluster like that of figure 2.8, we can see that the main sequence is unpopulated down to about 3 solar luminosities, the equivalent of

Figure 5.4 The H-R diagram of the globular cluster of figure 2.8, together with an evolutionary track (in black) indicating how the stars wander off the main sequence into the area of the red giants. Because of the difference in mass of the stars that have left the main sequence (stars in this cluster of the same mass as the sun are still on the main sequence), the somewhat different chemical composition of the stars in the cluster, and the difference between total and visual luminosity, the evolutionary track outlined here is quantitatively not comparable to the track of the sun outlined in figure 5.1; qualitatively, however, many stars in this cluster have attained a stage that the sun has yet to reach.

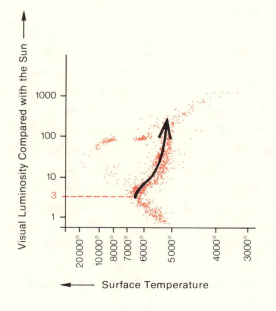

about 1.3 solar masses. That means that the brighter main-sequence stars of the cluster have already depleted the hydrogen at their center. Stars of more than 1.3 solar masses lie along a line branching off the main sequence toward the upper right into the area of the red giants. The evolution of these stars conforms pretty much to our calculations about the future of the sun, and with regard to their mass most of them also resemble the sun.

The black line in the H-R diagram in figure 5.4 traces the evolution of a sunlike star of that globular cluster. The stars in the cluster follow the evolutionary pattern predicted for the sun. We see stars move steeply upward to the right of the diagram, just as we expect the sun to do 8 billion years hence. Because they are in a more advanced stage of development these stars show us the future course of the sun. If any planets revolving around them at one time also sustained life, it has long since been extinguished by heat irradiation. The evolution of these stars alas confirms the validity of our predictions about the sun.

Solar Neutrinos

We have come up with a computer sun possessing the observed properties of our actual sun, and the H-R diagrams of globular clusters bear out the accuracy of our prognosis about the sun, however unpleasant that prospect for mankind. As far as astrophysicists are concerned, all's right with the world, except possibly for one discordant note, namely, the nagging doubt of nuclear physicists that the posited view of the life of the stars might not be altogether correct, that the computer models might be in error.

That doubt, supported by an experiment conducted in an abandoned gold mine in South Dakota, is connected with an insignificant elementary particle, a sort of by-product of the conversion of hydrogen into helium that basically has no effect on the sun. The particle in question is the neutrino, an electrically neutral object of practically no mass that moves with the speed of light. When we discussed the proton-proton chain we saw that when two hydrogen nuclei fuse, they emit a positron and a neutrino (see figure 3.3). The positron promptly unites with an electron, producing a photon, but the neutrino does not react with other particles and consequently, not being distracted, flies off from its birthplace with the speed of light. The neighboring solar matter does not affect the neutrino. As far as this newly created particle is concerned, solar matter simply does not exist. If we wanted to protect ourselves from an oncoming neutrino, we would have to take shelter behind a wall whose thickness expressed in kilometers would run to 15 digits. Fortunately, we do not have to protect ourselves from neutrinos, for they pass through our bodies without harming even a single atom.

Neutrinos originating at the center of the sun thus fly off straight into space and, regardless of the time of day, arrive at the earth's surface. During daytime they come down to us from above and at night they come from below passing through the earth with ease. If we had something like a neutrino telescope to make them visible to us, we would see a small bright speck in the center of the solar disk, the region in which proton-proton reaction takes place and where neutrinos are born. If this imaginary telescope were positioned below the horizon and aimed in the direction of the sun, it would also show the speck at night, long after the sun has set, for this telescope would be able to see through the earth.

However, we do not have neutrino telescopes. To make them we would have to be able to refract neutrinos with lenses or mirrors, as a camera refracts light or an electron microscope refracts electrons. But neutrinos in flight follow a straight path. Still, there are atoms that offer resistance, however insignificant, to the flying neutrinos. Among them—and the most well-known—is Cl^{37}, an isotope of the element chlorine. If any atoms can stop neutrinos those of this chlorine isotope are the most likely prospects. They rarely succeed in doing so, but when they do they swallow the incoming neutrino and in return emit an electron. The product of this trade is an atomic nucleus of the element argon (see figure 5.5). This argon atom, how-

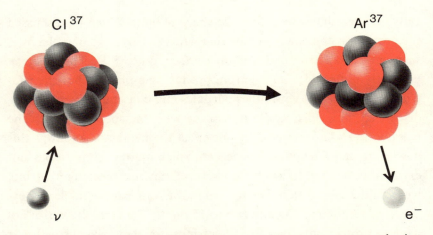

Figure 5.5 A neutrino can change a chlorine atom into an argon atom, releasing an electron in the process.

ever, is not the usual atom of this inert gas but an isotope that after about thirty-five days reverts to its former state. Raymond Davis of the Brookhaven National Laboratory conducted an experiment based on the interaction of the solar neutrino with Cl^{37}, and the results caused embarrassment among astrophysicists. But before going into Davis's experiment I would like to touch on some other points.

Chlorine atoms respond only to highly energetic neutrinos. Neutrinos originating in the sun's proton-proton chain are very low in energy and consequently have no effect on chlorine atoms. That in fact would put an end to our speculations about solar neutrinos if the sun did not possess another source of high-energy neutrinos. The proton-proton chain produces a series of side reactions that have not been mentioned earlier because they have no bearing on the energy production of the sun. One of these reactions increases in frequency with the increase in the amount of helium that has been formed (see figure 5.6). When a normal helium atom with the mass number 4 collides with a helium isotope with the mass number 3, the result is beryllium with the mass number 7. If this atom should meet a hydrogen nucleus before disintegrating radioactively, we would get a boron isotope with the mass number 8. This boron isotope, which is also radioactive, subsequently reverts to beryllium. In the course

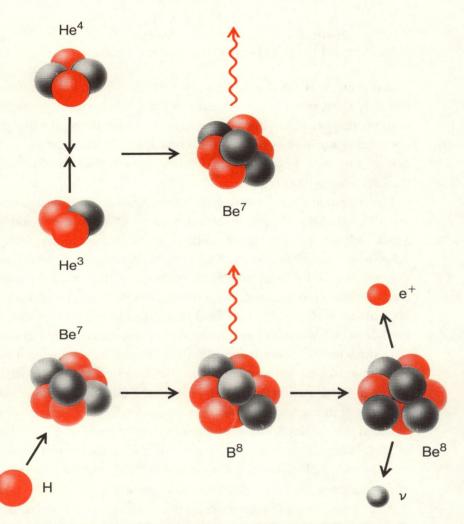

Figure 5.6 A side reaction of the proton-proton chain (see figure 3.3) produces the radioactive isotope Be^8, which emits one positron and one high-energy neutrino. Red, wavy arrows indicate the emission of photons.

of this reversion it releases a positron and a highly energetic neutrino.

These neutrinos are just right for a chlorine reaction. They, too, can penetrate matter, even great masses of chlorine, practically at will, although the chlorine atoms occasionally though rarely react to the neutrinos passing by. Their rare encounter formed the basis of Davis's experiment.

Raymond Davis's Neutrino Experiment

It is possible to build a detector for solar neutrinos, but it can detect only those neutrinos originating in the astrophysically speaking unimportant beryllium-boron side chain. It cannot detect the neutrinos originating in the proton-proton reaction so vital to the sun and hence to us. But if our solar models are correct, they should show the high-energy boron neutrinos.

Davis devised the following experiment: he buried a tank containing 390,000 liters of perchloroethylene, an industrial dry-cleaning agent related to the more widely know tetrachloride, in a 1,500-meter-deep excavation. The tank is surrounded by water to prevent any faulty reactions. Each molecule of this cleaning fluid contains four chlorine atoms, one of which is on average the neutrino-sensitive isotope Cl^{37}. This fluid offers the cheapest and simplest method of combining numerous chlorine atoms within a small space.

In the tank the atoms are constantly irradiated by solar neutrinos. For the most part nothing happens. The numerous neutrinos originating in the proton-proton chain pass through the tank unimpeded. Only the high-energy neutrinos that originated in the disintegration of boron run the danger of being captured. By estimating the number of high-energy neutrinos on the basis of the astrophysicists' solar models, we conclude that a solar neutrino would convert an average of one chlorine atom per day into an argon atom.

After a few days of exposure some argon atoms have formed. But the argon disintegrates after about thirty-five days and reverts to chlorine. Thus, if the fluid were exposed for any length of time to a stream of solar neutrinos, a sort of equilibrium would soon be established: on average the number of argon atoms that form equal the number that disintegrate. Unfortunately, the resultant concentration of argon atoms is very meager. If our solar model is accurate the tank should contain only about thirty-five argon atoms. These now have to be found and counted.

Finding 35 argon atoms in 610 tons of fluid makes finding a needle in a haystack look like child's play. The number of chlorine atoms alone in 1 cubic centimeter comes to a 22-digit figure, and Davis's tank contains 390,000 liters, or 390 million cubic centimeters. And we're supposed to find 35 argon atoms. Still, the problem can be solved. The argon atoms can be extracted by blowing helium through the fluid, a technique that has proved to be effective in experiments. By this method it is possible to fish out 95 percent of all argon atoms in the tank. Since the argon atoms originating with the solar neutrinos are radioactive, they are easily measured by Geiger counters once they are removed from the tank and disintegrate. In the fluid now free of argon atoms, new argon atoms can form, and they in turn can be extracted and counted. Perchloroethylene turns out to be an inexhaustible breeding ground of argon atoms.

On average one reaction per day is expected to take place in the tank. However, over the years tests have shown that on average a reaction takes place only once every four days. That would indicate that the sun's per second emission of high-energy neutrinos is only one-fourth as high as predicted.

The astrophysicists checked and rechecked their models while Davis continued to search for possible errors in his experiments. The discrepancy remained. Where had we gone astray in our calculations? Where is the error in the gold-mine experiment? It was hardly likely that all the computer answers were wrong. We knew that the computer models of the sun largely conformed to the actual sun, and since even minor corrections of the models would reduce the flow of high-energy solar neutrinos, the discrepancy between model and experiment could be eradicated. The only problem with these corrections is that they would reduce the temperatures at the centers of the solar models, and we can find no reason why these temperatures should be lower than those calculated by the computer.

A way out of the dilemma would be found if it were determined that the neutrinos were short-lived. Physicists studying elementary particles do not know too much about neutrinos. If like so many other particles they would disintegrate soon after they form, or dissolve into other particles during their eight-minute journey from the sun to the earth, it would not be surprising that the chlorine experiment yielded fewer than the expected number of neutrinos. But physicists think that neutrinos do not disintegrate on their way from the sun to us, which would seem to rule out this solution.

I for one do not believe that there are significant errors in the computer models. It is possible that the posited reaction rates of the beryllium-boron chain are in error. What would happen if the two helium nuclei, the ordinary helium and the lighter helium isotope, that stand at the beginning of this chain (see figure 5.6) interacted more rarely than atomic physicists have assumed? Would our sun then look any different? The answer is no, for this would not affect the proton-proton chain that delivers solar energy. Nothing about the sun would be any different except that the flow of high-energy neutrinos would be smaller, which would conform to the chlorine experiment. For that reason I do not believe that despite the chlorine experiment, any substantial revision of our idea of the internal structure of the sun is necessary.

The Gallium Experiment

Atoms other than chlorine also respond to neutrinos, among them an isotope of gallium. It has the mass number 71, and in combination with a neutrino it changes into the element germanium. The essential difference between the gallium and the chlorine experiments lies in the fact that the former also involves low-energy neutrinos. A gallium detector counts the neutrinos of the proton-proton chain, that is, the neutrinos released in the production of solar energy, rather than those of an unimportant side reaction.

So why has the gallium experiment not been performed? First of all there is the problem of counting the germanium originating with the neutrino reaction. Appropriate detectors have to be developed. Then we find ourselves in the same dilemma as in all other neutrino experiments. Only rarely do neutrinos let themselves be captured by atomic nuclei. Given the stream of solar neutrinos coming down to us day after day, it would take a tank filled with 37 tons of gallium to convert one gallium atom per day into an atom of germanium. That is no small amount compared with the total supply of pure gallium in the world. At today's market prices a ton of gallium, a by-product of aluminum, costs about half a million dollars. Of course, one might borrow the gallium needed for the experiment rather than purchase it, but it is questionable whether that would make for a substantial saving. There is no question that every major power is hoarding gallium, an essential product in the electronics industry, to cover its requirements in case of war.

At the time of this writing the Max Planck Institute of Nuclear Physics in Heidelberg was working on germanium detectors, and the United States, Israel, and West Germany were negotiating the acquisition of gallium for experimental purposes. At present they are talking about a ton for a pilot experiment. Will the experiment, which is sure to be carried out eventually, confirm our views of the internal structure of the sun? Or will everything we astrophysicists claim to know about solar energy production turn out to be wrong?

The reader may wonder why we have completely ignored some of the sun's properties in our preceding discussion. The reason we have neglected such phenomena as sun spots and their eleven-year cycles, prominences, and radiative explosions is our desire to concentrate on the more basic properties of the sun. Sun spots, prominences, and explosions are comparable to our weather. After all, an understanding of the evolution of the earth's interior does not necessarily involve an analysis of the phenomena of thunder and lightning.

6

The Life Story of Massive Stars

Astrophysicists have not been too upset by the still unexplained chlorine neutrino experiment. Other evidence shows that computer calculations of the evolution of the stars agree with observed celestial phenomena. These phenomena form the subject of this chapter, in which we will discuss the evolution of stars more massive than the sun. Because more massive stars use up their store of nuclear energy more rapidly, they are in a fairly advanced stage of depletion. They allow the astrophysicist to see whether computer projections about the later phases of stellar evolution conform to the cosmic reality.

The road to the computerized picture of stellar evolution was not without its problems. The big computers that came into use after World War II did not automatically produce better, more accurate results. Furthermore, the solution to the problem of stellar evolution called for new methods of calculation. The lay reader might be surprised to learn that not only computers but also new methods of calculation are essential to solve a numerical problem. It is easy to see that new types of telescopes or new space probes can be a boon to astronomers. But since new mathematical approaches cannot be made accessible visually in the form of wooden or plastic models or color transparencies nor unveiled ceremoniously before an admiring public, their role is less obvious.

Louis Henyey and the Henyey Method

After Hoyle and Schwarzschild's pioneering work of 1955, the theory of the evolution of sunlike stars, that is, of stars of about 1 solar mass, stagnated. The most highly evolved models that had been obtained described red giants [see figure 5.2(d)]. These models showed central temperatures of 100 million degrees. At that temperature helium fusion is supposed to begin. As we have seen in chapter 3, at this temperature helium atoms can fuse into carbon. Yet as soon as the models were asked to simulate the first nuclear reactions of this new source of energy, the process ran aground. That helium burning in these stars would set in fairly quickly and violently had already been demonstrated theoretically by Leon Mestel at Cambridge University in 1952, but no one suspected that the computer, using the then known methods, would refuse to construct models.

As far as massive stars were concerned, matters were still worse. Although it was possible to calculate the utilization of hydrogen in the convective central regions, the computers balked as soon as depletion effects became apparent. It was not even possible to follow the models of massive stars into the region of the red giants, something Hoyle and Schwarzschild had been able to do with sunlike stars. Nothing worked.

Although computer technology continued to improve the capabilities of its products, there was very little that astronomers could do with them. Hoyle and his colleagues tried numerically to follow the evolution of more massive stars without much success. Schwarzschild vainly tried to break through the helium-burning phase of sunlike stars. At the time a group of physicists in Japan was working with Chushiro Hayashi, who, using very simple desk calculators, sought to find out what happens to a massive star once the hydrogen in its center becomes depleted. As we later learned, when a new mathematical approach was found, Hayashi's work came closest to actuality.

The Henyey crater lies on the far side of the moon. It was given its name in 1970 by the International Astronomical Union in honor of Louis Henyey, who had died earlier that year. Henyey left his mark on many areas of astronomy, but the work for which he will undoubtedly be remembered best is a mathematical approach generally referred to as the Henyey Method.

In August 1961 The International Astronomical Union convened in Berkeley, California. It was the first such meeting I was privileged to attend. Some days prior to the meeting the Berlin Wall had gone up. Alfred Weigert, a young astronomer from Jena, East Germany, happened to be visiting relatives in West Berlin at the time and found himself faced with a dilemma: should he return home or stay in the West. He decided to stay, and before long he was to play an important role in the problem that forms the subject of this book. But back to Berkeley. Among the numerous papers read was one by Louis Henyey, a member of the astronomy department at the University of California at Berkeley, dealing with a new method of calculating stellar models. Rumors had been circulating for some time that Henyey had developed a new method. Some years earlier his team had already published a paper on this subject, but the approach it described was rather convoluted and no one, probably including Henyey himself, had managed to obtain useful results with it. But now the method was said to have been greatly simplified and improved.

Henyey was not given to voluminous or precipitous publication. That is why on that afternoon everyone working in the field of stellar evolution came to hear him. I did not understand anything he said, but I nonetheless took copious notes. When subsequently I spent six months working with Martin Schwarzschild at Princeton, I watched him reconstruct Henyey's process from notes. I thereupon went back to my notes and within a few days I understood how the Henyey Method worked. Schwarzschild promptly applied the method to the nagging problem of helium burning in sunlike stars. Before long he had worked through this rapid, almost explosive evolutionary phase. With the help of the Henyey Method, he was able to follow a star through a previously inaccessible phase of evolution.

I returned to Munich in the fall of 1962 after a brief visit to Pasadena, my elaborations of the Henyey Method safely tucked away in my briefcase.

In the interim Alfred Weigert had come to the Max Planck Institute after its relocation at Munich. He and a young actuary, Emmi Hofmeister, were ready to help me build stellar models according to the Henyey Method. The computer facilities at the Institute for Astrophysics, the successor to the Department of Astrophysics of the old institute at Göttingen, were exemplary, and we had clear sailing. We wanted to follow more massive stars from the main sequence into the region of the red giants, an area in which conventional methods had failed to work as soon as the models departed from the main sequence.

By March 1963 our star—we had decided on one of 7 solar masses—had not only left the main sequence but had long since grown into a red supergiant and begun to fuse helium into carbon. We sent a cable to Henyey at Berkeley that read as follows: "The Henyey Method is working in Munich. Thank you."

In those weeks was born the story of a 7-solar-mass star.

The Story of a 7-Solar-Mass Star

Why a star of 7 solar masses? The reason for our choice was our belief that in the course of its evolution such a star would pass through a stage in which it would possess all the properties of a certain type of variable star—the so-called *Delta Cephei stars.* Before this no one had ever observed the evolution of an ordinary main-sequence star into a Delta Cephei star. Now the effective Henyey Method held out that very possibility. In fact in the course of its evolution our star repeatedly passed through this stage. But before telling about this I would like to describe the step-by-step progress of a 7-solar-mass star.

Let us begin with the main-sequence stage. Here the star's interior is still composed uniformly of hydrogenous matter and possesses all

95

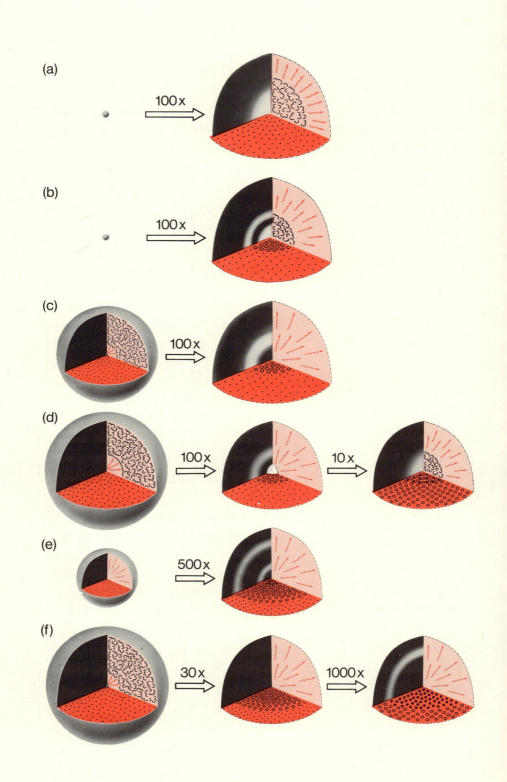

(a)

100x

(b)

100x

(c)

100x

(d)

100x   10x

(e)

500x

(f)

30x   1000x

the properties of a main-sequence star. What happens subsequently is illustrated in figures 6.1 and 6.2. Figure 6.1 shows the internal structure at different stages of evolution, beginning with the chemically homogenous first model of figure 6.1(a). The star's evolution is plotted in the H-R diagram of figure 6.2 together with the evolutionary paths of stars of different mass. It begins at the main sequence and, as anticipated, leads into the area of red supergiants. As we already know, the hydrogen supply of a star lasts a long time. We saw in figure 2.10 that a star of 7 solar masses can live off its hydrogen for 10 million years. In the course of this time span the amount of helium in the convective core increases substantially, yet the total structure of the star changes only imperceptibly. Its radius expands slightly, its surface temperature initially drops and then rises again, and its luminosity increases somewhat. Accordingly, the star in the H-R diagram (figure 6.2) at first moves slowly toward the right and then again to the left. During this entire period it remains on the main-sequence band. Approximately 26 million years after the burning of hydrogen begins, the nuclear energy stored in the core becomes depleted. Now the interior begins to change substantially. The cen-

---

Figure 6.1 The internal structure of a 7-solar-mass star at different stages of evolution. The models in the left column are drawn to the same scale. In those to the right the interior is enlarged. In those depicting later stages of evolution, the interior is doubly enlarged. The symbols used are the same as those in figures 4.2 and 5.2. After the ignition of helium, carbon forms, represented by small black solid circles. (a) The initial main-sequence model with a convective central region. (b) The star after 26 million years. Its diameter has not yet changed, but in the central region the change from central burning to shell burning, as shown in the upper left cross-section, has already begun. (c) 26.5 million years after the ignition of hydrogen, a helium sphere forms in the center. Hydrogen fusion now takes place in the shell. The star's radius has expanded. It has a thick outer convective zone, as shown in the upper right cross-section of the figure on the left. (d) 100,000 years later the helium has already ignited. Now the star lives off its hydrogen-burning shell and the helium fusion in the center. It has grown still bigger. (e) 34 million years after the hydrogen ignition, the helium in the center becomes depleted. The star now lives off two shell sources: the hydrogen-burning source in the outer layer, and the helium-burning source in the inner layer. The star shrinks temporarily and loses its outer convective zone. (f) 2 million years later it once more becomes a red supergiant. Its hydrogen-burning layer is temporarily extinguished, and the star lives off helium fusion. Its chemical composition has become quite complicated. On the surface we still find the original hydrogen-rich matter; beneath it a heavy layer of helium surrounds a minute central sphere of carbon.

tral region no longer produces the energy required for the star's luminosity. Consequently, the hydrogen somewhat farther away from the center, a shell-like area surrounding the spent core, begins to burn. As in the evolution of the sun, a shell source forms [see figure 6.1(b)]. Only helium fills the interior of the shell source. The star now has a helium core on whose surface hydrogen is converted into helium, and only farther outward can the original hydrogenous matter still be found.

Now things happen quickly. The helium core within the shell source contracts and heats up while the outer layer of the star expands and cools off. The surface temperature drops considerably although its luminosity continues unchanged. In the H-R diagram the star moves horizontally toward the right. It turns into a red supergiant [see figures 6.1(c) and 6.2]. This transformation takes a mere 500,000 years. In this relatively brief period the star traverses the entire H-R diagram from left to right.

Figure 6.2 The evolution of stars with different masses. The figures along the track indicate the respective mass in solar-mass units. Whereas the evolution of a 1-solar-mass star—as we know from figure 5.1—leads into the region of the red giants, the track of more massive stars leads to still bigger red stars, the supergiants. The symbols along the red tracks of the 7-solar-mass stars refer to the models of figure 6.1. The two parallel broken straight lines define the strip in which the Delta Cephei stars are found.

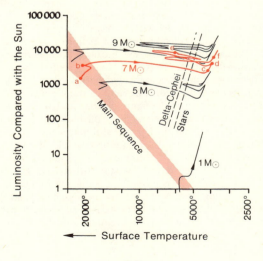

In the region of the red supergiants, a new phenomenon now appears. During the cooling off period the outer layers lose some of their radiative transparency and convection takes over the transport of energy. The star has acquired a thick outer convective zone, reaching down from the outside into the innermost region. For a time 70 percent of the entire stellar mass is located in this outer convective area. But the zone of ascending and descending matter does not penetrate deeply enough to commingle the helium originating in the central region with the outer matter. The helium remains untouched in the vicinity of the center.

But the interior also enters into a new phase of evolution. While the outer area was expanding the burned-out helium core contracted sharply. The center density increased so greatly that now a cubic centimeter contains 6 kilograms. This compressed matter also grew hotter until it reached a temperature of 100 million degrees. We already know that at such temperatures helium can change into carbon; after 26.5 million years of hydrogen fusion our main-sequence star opens up a new source of energy for itself, converting helium into carbon (see figure 3.4). As with the burning of hydrogen helium burning is concentrated in the innermost section of the star near the center, and again a new though comparatively small convective zone comes into being. Two energy sources combine to meet the star's luminosity requirements: the shell source in which hydrogen changes into helium, and the nuclear reactions in the center, where helium is transformed into carbon [see figure 6.1(d)].

The evolution of our stellar model, and with it the chemical structure of our star, now becomes quite complicated. Carbon begins to become enriched in the central core, and in the course of time the helium stored there becomes depleted. Six million years after its ignition the helium in the center is exhausted. Once again a shell source forms in which helium is converted into carbon. On the outside we find the original hydrogenous mix the star had at birth; below it, a layer of helium; and imbedded in that layer, a carbon core. Nuclear reactions take place both where the original mix borders on helium and changes into it, and further inside, where helium borders on the carbon in the central core and changes into it. The star now possesses two shell sources [figure 6.1(e)]. In the H-R diagram it moves to and fro, although for the most part it remains in the area of the red giants. Soon the outer shell source becomes extinguished, whereupon helium fusion becomes the star's sole source of energy [figure 6.1(f)]. The subsequent processes are more complicated still. Sooner or later the central region reaches temperatures at which carbon changes into other elements, and nuclear burning continues.

This is how in 1963 we reconstructed the story of a 7-solar-mass star. Since then many others, among them Pierre Demarque and Icko Iben in the United States, have carried out similar calculations for stars of other masses. Icko Iben in particular, working at the University of Illinois, tracked every detail in stellar nuclear reactions, and he has continued to investigate the mechanism by which newly formed elementary isotopes surface, for the atmosphere of a number of stars contain chemical elements that must have formed only recently in the stellar interior. The Polish astrophysicist Bohdan Paczynski is also doing pioneering research. In 1965 he was working under a handicap, for his computer in Warsaw was considerably smaller than those available to his colleagues in other countries; he nonetheless succeeded in developing a program based on the Henyey Method. We are indebted to him for many important findings, particularly on the evolution of binary systems.

But to return to our star. On the whole it is safe to say that stars ranging between 2 and 60 solar masses behave pretty much the same as our 7-solar-mass star. Stars of less than 2 solar masses develop much like the sun.

Evolutionary Paths and Star Cluster Diagrams

We do not know exactly whether a star evolves as predicted by computer calculations. However, the aspects we have already examined allow us to make a comparison between our model and observed phenomena and to test whether what our computer had to say about the evolution of the stellar interior is borne out by observation. Unfortunately, given the time spans involved, we cannot observe directly whether changes in luminosity and surface temperature conform to the theoretical path leading from the main sequence to the region of the red giants. It is therefore necessary to test the validity of the theory by other, indirect methods. Let us take another look at figure 6.2 and at the evolution of stars of 1 and 7 solar masses. Both eventually move from the main sequence into the area of red giants and supergiants. Let us assume that hydrogen fusion sets in at the same time in both. After some million years the more massive star would move to the right while the less massive one would remain on the main sequence for some billion years.

When we look at star clusters we see stars of various mass. The more massive ones are in a more advanced stage of evolution than their less massive neighbors, even though their age is the same. To illustrate this point, in the 1960s Alfred Weigert and I devised a method of demonstrating the different pace of evolution of different types of stars within a star cluster. We imagined an artificial cluster of 190 stars with solar masses ranging from 25 down to 0.5. The frequency distribution of these stars over the different masses was assumed to duplicate that found in an actual cluster. Thus, only 6 stars had more than 10 solar masses, while 42 ranged between 1 and 2 solar masses. The evolution of each of these stars was calculated.

Let us begin at a point in time at which all stars lie along the main sequence and plot the H-R diagram of this imaginary cluster. The result is a completely normal main sequence [figure 6.3(a)]. After only 3 million years the brightest main-sequence star, which we know to be the most massive one, shows the first sign of exhaustion: it leaves the main sequence. Thirty million years after the ignition of hydrogen several massive stars have moved to the right [figure 6.3(b)]. Some of the stars in our cluster, namely, the most massive ones, have already gone through the now familiar stages of stellar evolution and have arrived at a stage with which theory has not yet caught up. These stars have been omitted from the diagrammatic representation.

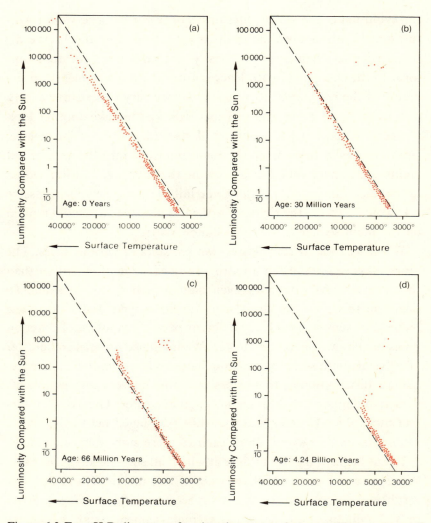

Figure 6.3 Four H-R diagrams of an imaginary star cluster at different stages of its life. The dots stand for stars of different masses, and in the course of time they follow the evolutionary tracks as predicted by computer models. The four diagrams show where the dots are located at four different stages of evolution.

The H-R diagram of the stars aged 30 million years also shows various characteristics of observed H-R diagrams. The main sequence is populated only up to a certain luminosity, and red supergiants are found at the right. Figure 6.3(c) shows the imaginary star cluster 66 million years after the onset of hydrogen burning. The main sequence is depopulated further down; again, some stars, this time of somewhat smaller mass, have wandered into the area of the red giants.

Figure 6.3(d) shows the H-R diagram of our imaginary star cluster at the ripe old age of 4.24 billion years. As we see it has undergone considerable change. Stars are concentrated in the lower portion of the main sequence, in a line branching off the main sequence to the right, and in a steeply ascending line on the right. This diagram differs from those preceding because stars of lesser mass follow a different evolutionary path; we now find that we are dealing with sunlike stars in the area of the red giants. The shape of this diagram is typical of very old star clusters, as we can see by comparing it with the H-R diagram of the globular star cluster of figure 2.8. This comparison also makes evident a limitation of our model calculations. We find that in both diagrams the lower portion of the main sequence is populated by stars and that the stars cluster along a line curving to the right and upwards, but we also find many stars clustering along an almost horizontal strip whose luminosities in the visible light are 100 times that of the sun. This so-called *horizontal branch* of the globular cluster's H-R diagram is not found in the H-R diagram of our imaginary cluster. As we pointed out earlier, stars in an actual cluster may go through evolutionary phases with which theory has not yet caught up. The diagram of the imaginary star cluster does not include stars that have gone through the known evolutionary phases. They are now missing from the diagram. We will return to the H-R diagram of globular clusters in chapter 7.

Pulsating Stars

Let us return to the evolution of a 7-solar-mass star. We have not yet mentioned that our star repeatedly intersects a particularly interesting strip in the H-R diagram, indicated by two broken parallel lines in figure 6.2. This strip is the home of all Delta Cephei variables.

Delta Cephei is one of the bright stars in the constellation Ce-

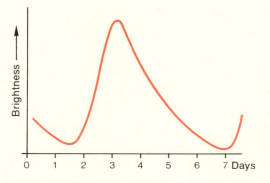

Figure 6.4 The light curve of Delta Cephei. In a 5.4-day cycle the star attains its maximum luminosity and then grows fainter again.

---

pheus. In 1784 John Goodricke—we shall come back to another important discovery of this prematurely deceased deaf-mute English scientist—observed that this star varied in brightness. Within a regular five-day cycle the star brightened and then dimmed again (see figure 6.4). At maximum brightness it was 2.5 times as bright as at its faintest level. Since then many such stars have been discovered. The periodicity of their light variation ranges between one and forty days. They have a surface temperature of around 5,300°, and their luminosity indicates that they are mature rather than main-sequence stars: they are red supergiants.

In the course of its evolution a 7-solar-mass star frequently goes through this stage. First it traverses the Delta Cephei strip from left to right and then from right to left, this last trip takes 350,000 years. The helium in its interior has long since ignited, and the star moves slowly, propelled by the nuclear burning of its helium. What happens to a star when it crosses the Delta Cephei strip? Why does the luminosity of a star within the strip vary? What is responsible for the periodicity of this variation? Today we know that the star's luminosity is not the only thing that changes; the star itself expands and contracts in rhythm with its light variation. It pulsates. Why do stars pulsate when they find themselves in a particular strip of the H-R diagram?

The answer to this question can already be found in Eddington's 1926 book on the internal structure of stars. Yet at his death in 1944 Eddington did not know how close he had come to the solution of the problem. In 1952 the Soviet mathematician Sergei Zhevakhin, following Eddington's lead, tackled the problem again. At first his work attracted little attention. But in 1960–61 John Cox (now at the Joint Institute for Laboratory Astrophysics at Boulder, Colorado) and, at the same time working jointly in Munich, Norman Baker (now at Columbia University) and I were able to prove that the Eddington-Zhevakhin theory held the explanation for the pulsation of Delta Cephei stars. Even though we are still a long way from understanding all the properties of these variable stars, generally speaking we know why they pulsate. I will try to demonstrate why with the help of a simple model, although this sort of model can give only a most elementary outline.

A Piston Model of a Delta Cephei Star

Gravitation is the force holding a star together. In an ordinary star gravitation and gas pressure are, as pointed out before, in equilibrium. This equilibrating property can be demonstrated by a simple model. Figure 6.5(a) shows how a movable heavy piston seals a cylinder at the top. The cylinder contains gas, which is compressed by the piston and prevented from escaping. Even though gravitational force tries to pull the piston down, it does not succeed altogether. The piston comes to rest at a certain distance from the bottom of the cylinder. If it were to continue its descent, it would compress the gas too greatly, and the increased gas pressure would push it back to its position of rest, or equilibrium. When the piston is at rest the gravitational force exerted on it and the counterforce of the gas pressure are in equilibrium. This position is analogous to the equilibrium of gravitational force and gas pressure throughout the interior of a star.

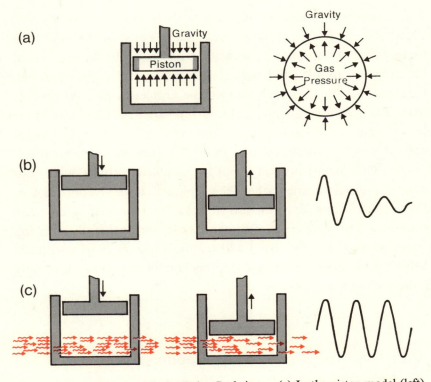

Figure 6.5 The piston model of a Delta Cephei star. (a) In the piston model (left) and in the star (right), gravity and gas pressure are in equilibrium. (b) The piston, set in motion, comes to a stop after a few oscillations because of the losses caused by friction. (c) Radiation permeates the gas in the piston model. If the gas in its compressed state absorbs more radiation than in its expanded stage, the piston can be kept in motion despite the friction.

If, however, the piston is pushed down from its rest position and suddenly released, it begins to oscillate. If it comes to a stop below its equilibrium position, the gas pressure will be too great for its weight, and it will be pushed up. If it comes to a stop above its equilibrium position, the gas pressure will be too low and gravitation will pull the piston down again. The piston does not come to rest at its equilibrium position, for once in motion its inertia causes it to overshoot the equilibrium position, and it oscillates between the two extremes. Set in motion the piston oscillates around a middle position. The gas acts like a spring. While expanding, it returns the energy it has received from the piston in its compressed state, and the piston in turn gives off its energy to the gas during compression. In our model friction may be negligible. No energy is ever lost. The piston oscillates with a certain period. The oscillation is not attenuated; the maximal deviation of the piston from the middle position remains constant. The period of oscillation depends on the properties of the model, that is, the mass of the piston and the median temperature and median density of the gas.

Stars behave in roughly the same fashion. If a star were to be compressed with equal force from all sides and then decompressed, the increased gas pressure would push the stellar matter uniformly outward, and it would overshoot its equilibrium position. But then gravity would exceed the gas pressure and pull the gas back toward the center. The star would pulsate. Once the star has lost its equilibrium in this way it continues to oscillate. Its oscillating rhythm, like that of our piston model, can be calculated if its properties—its mass and interior density, or in other words its internal constitution—are known.

The preceding description of both piston model and star is, however, overly simplified. The piston suffers a natural loss of friction. Once set in motion its oscillations taper off, become muted, and after a time it comes to a stop [see figure 6.5(b)]. In the case of a star it is not so much friction as other mechanism that impede oscillation. A star set in motion artificially will most likely come to rest between about 5,000 and 10,000 pulsations, after about 100 years. Yet we know from observation that Delta Cephei since its discovery in 1784 has been pulsating continuously. What motor force keeps it in oscillation even though according to our calculations it should die down in a relatively short time?

In his book Eddington mentions a possible mechanism for the oscillation of Delta Cephei stars. Powerful radiation from their center permeates the outer layers of all stars. To simulate this radiation in our piston model, let us imagine that the cylinder is made of transparent material and permeated by radiation moving from left to right [see figure 6.5(c)]. The gas in the cylinder, like stellar gas, is not completely transparent. It absorbs some of the radiation. Consequently, it will initially heat up until the difference in temperature between the gas and the surrounding atmosphere is so great that the cylinder will give off as much energy per second as it receives through its partial absorption of radiation.

We began with the piston in an equilibrium position. We now push down on it a little. The gas is compressed, and both pressure and temperature increase. In principle two possibilities exist now. At the moment of maximal compression the gas absorbs either more or less radiation. Let us take the first of the two possibilities. If absorption increases with compression more radiation is absorbed when the piston is pressed down than when it is in its resting position. This supplementary energy heats up the gas. The pressure increases and pushes the piston up powerfully to above its original position of rest. But here the gas is thinner and cooler, and consequently absorbs less energy. The gas cools off, its pressure diminishes, the piston moves down again, and that keeps it in motion despite friction.

What's true for the piston model is true for the star. If matter in one layer of a star also has the ability when compressed to absorb more radiation and transform it into heat, the radiation that permeates the star can set off oscillations. For when the star is compressed the radiation emanating from its interior toward the outside cannot penetrate the outer layers as readily. The gas then heats up and causes the star to expand. At the point of maximum expansion the stellar matter becomes less opaque. More energy can escape to the outside, the interior cools off, and the star contracts again; expansion is followed by renewed compression. Stellar matter affects outward radiation like a valve that opens and closes in rhythm with the star's pulsation.

Eddington had already described this mechanism in his pioneering work, and now comes the tragic part. Unfortunately, at the time little was known about the opacity of stellar gas, and what was known seemed to indicate that compression made stellar matter less rather than more opaque. In that event the effects would be contrary to those we have just described. The absorptive mechanism would dampen rather than stimulate oscillation. That was why Eddington himself put aside his mechanism and never let up in his search for another explanation for the pulsation of Delta Cephei stars.

Zhevakhin's New Approach to an Old Idea

By the beginning of the 1950s the opacity of stellar matter had been researched more thoroughly. Eddington's ideas about the stellar interior were proved correct. But it became clear that in some situations matter in the outer layers of stars could lose transparency under compression. That is precisely what happens at surface temperatures of around 5,300°. In 1953 Zhevakhin in a basic but long neglected work showed that in the case of a Delta Cephei star the effects of opacity properties of surface regions on oscillation were sufficiently strong to overcome the damping effect in the rest of the star and to allow the star to oscillate. Thus, Eddington's radiative valve mechanism can overcome the damping effect and keep a Delta Cephei star oscillating.

After our Munich team in 1963 found that the evolutionary path of the 7-solar-mass star transversed the Delta Cephei strip 5 times, it was only logical for Norman Baker and me to resume the work we had begun in Munich in 1960—calculations to test the oscillating properties of stars. We found that every time our model crossed the Delta Cephei strip it started to pulsate, and the periodicity of its oscillations agreed with observed periods. Thus, we can conclude that Delta Cephei stars and their oscillating properties conform to the scheme of stellar evolution. When a star in the course of its evolution crosses the Delta Cephei strip in the H-R diagram it oscillates, and when it leaves the strip the stimulative mechanism in the outer layers is no longer strong enough to keep the star oscillating.

Martin Schwarzschild once put it thus: Delta Cephei stars are like people with the measles. When they have them everybody can see it, but later, once it's all over, there is no way of telling whether or not they had ever had them.

7

Highly Evolved Stars

What happens to our 7-solar-mass star once the helium in its center has been used up? Do things continue from energy crisis to energy crisis? Does the core continue to heat up until it reaches 300 million degrees and the carbon ignites? When it does reach this point it becomes difficult to follow the star on the computer. After the helium in its center is exhausted, density and temperature increase; everything moves toward carbon burning. And then problems arise.

Neutrinos Cool; Shell Sources Flicker

When density and temperature in the stellar interior have become sufficiently high and a photon and electron meet, two new elementary particles are formed (see figure 7.1). We have met one of these, the neutrino, before; the second one, the *antineutrino,* a close relative of the neutrino, resembles it in many ways. Specifically, it too can go through stellar matter unimpeded. Stars are thus transparent both to neutrinos and antineutrinos. The birth of the neutrino-antineutrino pair involves an expenditure of energy by their parents, the electron and photon. This energy is contained in the newborn twins and escapes with them from the star's interior into the cosmos. When the star's center contracts and reaches the temperature needed to burn carbon, neutrino-antineutrino pairs continue to form, carrying

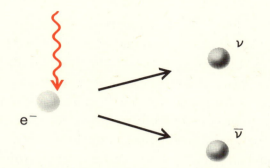

Figure 7.1 At temperatures of 100 million degrees a neutrino and an antineutrino can form when an electron (gray sphere) and a photon (red, wavy arrow) meet.

off energy and cooling the stellar interior, thus hindering, or at least delaying, carbon ignition. When carbon fusion finally does take place, the delayed reaction carries explosive force, possibly even shattering the star. If we want more precise data on this phenomenon, we would have to carry the calculation up to this phase. But here new problems crop up.

In the advanced stages of a star's evolution in which energy is produced in hydrogen- and helium-burning shells, nuclear reactions are no longer uniform. Over the course of centuries energy production increases and ebbs. At times, the star's luminosity is fed by its hydrogen-burning shell, and at other times the helium-burning shell takes on the job of energy production. Convective regions form above the energy-producing shells, intermingle with the matter above the shells, and disappear again. If one wanted to follow these processes in more detail on the computer, one would have to track the flare-up and dying down of the two energy-producing shells in small time steps, and it would probably require hundreds of stellar models to replicate a time span of even a hundred years of the actual life of a star. Anyone who wants to follow the development of a star over millions of years is therefore facing a well-nigh impossible task. To date no one has succeeded.

Even if we had managed to solve this problem, others would surely come up. Nuclear burning becomes increasingly complicated. When two atomic carbon nuclei collide and react with each other, the end product is by no means predictable. It could be magnesium or nitrogen, neon or sodium. All these atoms are formed in specific abundance ratios. The chemical structure of the star becomes more and more complex. Furthermore, because the fusion of different higher elements takes place at almost identical temperatures, a number of different nuclear reactions can take place almost simultaneously at any given site in the star. In the face of these phenomena the builders of stellar models have for the time being thrown in the towel. Here the art of simulating the life story of a star with electronic computers has reached its limits. We do not know how the story continues; all we can do is guess.

The White Dwarf in the Red Giant

Since the computer apparently is unable to tell us what happens next, perhaps observation can provide us with some clues. What should we be on the lookout for in the sky? What do the latest stages of models which we have managed to squeeze out of the computers look like?

During the evolution of our 7-solar-mass model from main sequence to final recorded phase, its central region continues to contract. Density increases considerably, first after the depletion of hydrogen and then again after the depletion of the helium at the center. While the star is still living on the main sequence, its central density amounts to less than a tenth the density of water, but as the helium in the center becomes depleted, it becomes increasingly denser, finally reaching a density of 10 tons per cubic centimeter. The only other known examples of such material density is found in white dwarfs.

The interior of our developed star does in fact contain an extremely dense core whose mass is slightly more than 1 solar mass and whose radius is identical to that of a white dwarf. It bears a striking resemblance to a white dwarf, except for one thing—it is surrounded by a gigantic shroud of gas of about 6 solar masses. That, incidentally, is typical of all red giants and of the even brighter super-giants that have already used up the helium at their center. They all, like our 7-solar-mass star, have dense cores. A white dwarf lies hidden in the heart of every red giant. If we could lift off the shroud that surrounds the dense core, we would find hidden beneath a star that in no way differs from the white dwarfs that inhabit the heavens. Can a developed star shed its outer layer and transform itself into a white dwarf, into a star like the Sirius companion?

Before we pursue this question any further, let us briefly leave the more massive stars and turn our attention once more to sunlike stars. How far can our model follow them?

The Further Future of the Sun

We pointed out earlier that the rapid onset of helium burning posed a problem for the computer simulation of the evolution of sun-like stars. However, using the Henyey Method, Schwarzschild and his co-worker Richard Härm succeeded in following the rapid burning of helium, the so-called *helium flash*. What is taking place in the star? The following elaborations are based on Hans-Christoph Thomas's dissertation, presented in Munich in 1967.

To recapitulate: our sunlike star is found in the upper right of the H-R diagram (see figure 5.4); it has long ago exhausted the hydrogen at its center. There a helium core has formed on whose surface hydrogen continues to burn, eroding the still hydrogen-rich shell. The shell itself is greatly enlarged, for in the interim the star has turned into a red giant [see figure 5.2(d)].

While the helium core continually increases in mass because the hydrogen on its surface is being converted into helium, both its central density and temperature increase. Soon here, too, photons and electrons begin to produce neutrino pairs, and some of the internal energy escapes with the neutrinos. The central region is cooled by their departure. Although stellar temperatures are generally highest

113

at the center, in this instance the central temperature, because of the cooling effect of the escaping neutrinos, is lower than a layer that is still part of the helium core even though it is a considerable distance from the center. In this, the hottest region, the helium soon ignites. Since the helium fuses at high density, it flares up spectacularly, producing the helium flash. Helium burning in this case is very rapid, but the sun, once it reaches that stage, will look no different to the outside observer. Externally, the inert solar body will react only slightly to the fleeting intensified energy production in its interior. Its helium will burn rather vehemently for about two thousand years and will be followed by a steady nuclear fire.

At that point all the geriatric ailments of aged stars resurface. The shell-burning sources flicker and repeatedly force the computer to deal with processes that take place within a one-hundred-year span. It becomes impossible to cover time spans of millions of years, the actual intervals during which the star evolves.

With that we seem to have arrived at an impasse. We now ask ourselves whether we are perhaps looking at stars that have already passed through this phase and therefore are able to tell us what the future holds in store. Here the H-R diagram of the globular cluster of figure 2.8 comes to our assistance. In it we saw stars moving from the main sequence to the realm of the red giants. We know that these are stars that have not yet burned their helium. Since our calculations show that a star remains in the upper right of the diagram while it is igniting helium, we must conclude that the stars that form the horizontal branch of the diagram have already begun to burn helium. But the computer models showing stars after the helium flash do not have them gravitating toward the horizontal branch on the left. They stay on the right, in the area of the red giants. How then do the stars in the universe get to the horizontal branch?

John Faulkner, a student of Hoyle who is now at the University of California at Santa Cruz, was the first to tackle this problem. He showed that it was possible to conduct a little experiment with the computer models of helium-burning sunlike stars. If we subtract some mass from their surface and let the computer figure out the

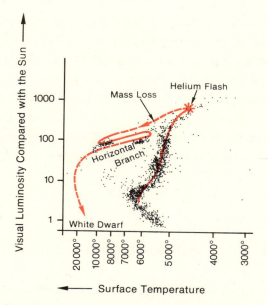

Figure 7.2 A schematic drawing of the evolutionary track of a sunlike star in the H-R diagram. After its main-sequence stage the star turns into a red giant, as shown in figure 5.4. There its helium ignites (helium flash). As a red giant it blows off so much matter from its surface that it loses a substantial portion of its outer layer and lands on the horizontal branch of the diagram. After that it most likely turns into a white dwarf. For purposes of comparison the stars of the globular cluster M3 of figure 2.8 have been inserted into this illustration.

structure of these partial amputees, we find that these stars are no longer found on the right of the H-R diagram but near the horizontal branch. It is not necessary to remove all of the hydrogen shell surrounding the helium core; a partial amputation suffices. Has this computer experiment brought us closer to the truth? Do sunlike stars perhaps lose mass from their surface in their red-giant stage, and do they then, freed of part of their shroud, settle on the horizontal branch of the H-R diagram, where the diagram of the globular clusters show stars whose helium apparently has already ignited? Let us look at figure 7.2. Does it reveal what the future holds in store for our sun? Will it as a red giant lose so much mass as significant portions of its outer layer are blown into space that it will remain on the horizontal branch of the H-R for a longer period? That seems probable. Sooner or later almost the entire solar mass will be concentrated in its white-dwarf core, and ultimately—perhaps after a phase during which it throws off the remainder of its outer layer—the sun itself will turn into a white dwarf.

Now that the computer models of highly developed stars show us that stars have to lose mass, let us look for specific indicators in the sky. A great many of them can be found, not only among the highly developed stars but also in so harmless a main-sequence body as our sun.

Peter Apianus, Ludwig Biermann, and the Comets

Peter Apianus (the pseudonym of the sixteenth-century German mathematician Peter Bienewitz) taught astronomy at Ingolstadt, Bavaria. Ludwig Biermann, who lives in Munich, was my predecessor at our Max Planck Institute. The story I am about to tell deals with the peculiar properties of comets and ends with the question about the sun's loss of mass.

Comets are bodies composed of less than a millionth part of the earth's mass. They move around the sun in elongated elliptical orbits. The best-known of them, Halley's comet, has an orbital cycle of about seventy-five years and will be close to the sun again in 1986. When comets enter the vicinity of the sun, gas masses, which probably emanate from cometary ice or snow, evaporate. With them dust particles that had been mixed in with the snow are also released. The cometary gas and dust particles do not fly out symmetrically in all directions. Rather, they form the characteristic cometary tail. To be precise two tails emanate from the comet, one of dust along which the dust particles fly off, and one of gas. The pressure of the sun's light moves the dust particles into a frequently curved orbit away from the sun. For the time being we are not interested in the dust tail but in the gas particles. They fly away from the comet at high rates of speed along a straight line, occasionally reaching 100 kilometers per second.

The striking appearance of the comets (not to be confused with the meteors flitting through the sky) has stirred the imagination and aroused the curiosity of mankind since time immemorial (see figure 7.3). In the Middle Ages they were thought to be omens of ill fortune—war, famine, and pestilences—but they have always stimulated scientific thinking as well. In the early sixteenth century Apianus discovered that the luminous cometary gas tail was always turned away from the sun. A comet traveling through space does not leave its gas tail behind in its orbit. Whatever its course its gas tail always points away from the sun (see figure 7.4). When the comet moves away from the sun its gas tail precedes it. The turning of the tail away from the sun and the high velocity of the gas pouring out of the comet as it moves away from the sun gave rise in the nineteenth century to the idea that a counterforce to gravity was repelling matter from the sun. The only known force that would have qualified for that role was the pressure exerted by the light of the sun on the particles of the cometary tail, but in 1943 the astronomer Karl Wurm (1899–1975) in Hamburg demonstrated that radiative pressure was far too weak to account for velocities in the cometary tail.

Yet the high speeds had been observed and begged for an explanation. Since the gas particles always flew away from the sun, the sun had to be responsible. It occurred to Ludwig Biermann in 1950 that perhaps particles constantly flow through our solar system, carrying with them evaporating molecules from the comet's nucleus. We already knew that during occasional outbreaks on the sun, gas clouds are thrown into space, which explain such phenomena as the aurorae. But Biermann asserted that independent of solar outbreaks a constant wind of charged particles emanated from the sun. These charged particles—primarily hydrogen nuclei—carried the charged particles of the gases released by the cometary nucleus, while the uncharged molecules remained in the cometary head. The solar wind postulated by Biermann as the explanation for the direction of the cometary tail has since been confirmed by satellites, and space probes have measured its force and direction.

This answers the question posed by Apianus's discovery that the cometary gas tail was always directed away from the sun.

117

Figure 7.3 The comet Mrkos of 1957, with the straight gas tail pointing away from the sun, and the diffuse dust tail curving toward the left. (Palomar Observatory Photograph)

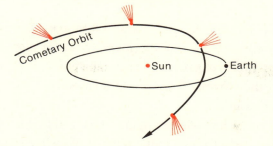

Figure 7.4 While a comet moves in its orbit the gas tail always points away from the sun.

The sun continues to lose matter. Does that mean that all calculations based on the assumption that a star retains its original mass are wrong? Does this finding perhaps offer a way out of the still unsolved neutrino paradox?

We now know that the sun sends forth 10 trillion tons of matter per year in the form of solar winds. Despite this huge amount, the sun in its billions of years of existence has not lost a significant amount of its mass. In its main-sequence phase the sun seems to be holding on to its mass, even though it gives off gas in which cometary tails flutter like flags in a breeze.

Developed Stars Lose Mass

While the sun loses hardly any mass during its main-sequence phase, developed stars lose greater amounts. Even if we do not yet understand the operative mechanism, we know that many red giants release gas from their surfaces into space. We have not yet been able to find the cause for the solar wind, but the velocity of the matter flowing from stars can still be measured and the loss of stellar mass can be estimated. Accordingly, some stars shed mass 10 million times more rapidly than does the sun. As a matter of fact in most instances the loss of mass is so substantial that a considerable portion of a star's mass is dispersed into space within a period of 100 million years.

Red giants are not the only stars to lose mass; hot, massive stars that have only recently left the main sequence also emit gas into space. Their wind velocities are particularly great. It is not unusual to find that they throw off matter at a speed of 2,000 to 3,000 kilometers per second.

The far greater loss of mass by developed stars does not, however, mean that we have to revise all our ideas of stellar evolution. As far as developed massive stars are concerned, 100 million years is still a long time, far longer than the phase of the ignition and depletion of helium in the central region. The loss of mass is significant only in the case of sunlike stars that have turned into red giants. But this in particular enables us to explain the stars along the horizontal branch of the globular star clusters.

Before I proceed I would like to mention another example of the mass loss sustained by developed stars. It involves the red giant Mira of the constellation Cetus. In 1596 the German theologian David Fabricius discovered that Mira was visible only occasionally and then disappeared again for extended periods. Today we know that Mira changes in brightness in eleven-month cycles. At its minimum level Mira is one six-hundredth as bright as at its maximum. Many red giants, for reasons still not known to us, undergo such swings in brightness. We do know, however, that the mechanism responsible for these swings is not the same as that of the Delta Cephei stars. But what interests us here is not the variability of this developed star but the behavior of its companion. When Mira is at minimum brightness, a companion star, a white dwarf, becomes visible, which at all other times is outshone by the bright red giant. As we know, a white dwarf revolves around Sirius. The companion of Mira takes 261 years to complete its orbit around the giant.

The South African astronomer Brian Warner found that the light of the white dwarf flickered. White stars are inert, barely variable stars. What, then, is responsible for the restlessness of the Mira companion? If Mira, Warner maintained, releases matter into space like most red giants, its companion moves in Mira's stellar wind and its gravitational force pulls some of the outflowing gas to its surface.

The gas collides with the companion star at a high rate of speed because of the companion's powerful gravitational force. The collision generates heat, and the hot, onrushing gas contributes substantially to the dwarf star's luminosity. Fluctuations in the flow of the onrushing gas produce the variations in the white dwarf's luminosity. Brian Warner did not need to find a particularly high loss of mass from Mira to explain the luminosity of the white dwarf and its fluctuation. In this instance, too, the loss of mass seems to have comparatively little bearing on Mira's evolution.

The observed rates of stellar mass loss thus help explain how sunlike stars reach the horizontal branch of the H-R diagram, but they do not help answer the question whether massive stars in the course of their evolution shed so much matter that one is left with a white dwarf, which was the question that we posed at the outset. Fortunately, we know of a celestial phenomenon that indicates that stars can give off greater amounts of mass within shorter time spans.

The White Dwarf Is Unveiled

Even with a small telescope it is possible, if one knows where to look, to see a small luminous ring in the constellation Lyra, the so-called *Ring nebula of Lyra*. We know of seven hundred such formations. Because to the observer they look almost like luminous disks, like the disks of planets, they are called *planetary nebulae* (see plate IV). But they are not related to the planets of our solar system. Located way out in space, these luminous nebulae are in actuality hollow spherical shells of gas, with a hot star in the center of each. The radiation from the hot star makes the nebula glow. One can see the gas nebula expanding at rates of about 50 kilometers per second. This expansion seems to indicate that the gas has been suddenly thrust off by the star. The luminous matter of the nebular shroud contains between 1/10th and 1/5th of a solar mass, not an insignificant amount compared with the mass of a star.

We do not know why the star thrusts off matter nor do we know the mechanism responsible for this loss of mass. All we know is that we have seen it happen. And we can see still more. A closer examination of the central star shows that its properties resemble those of a white dwarf: the surface temperature is very high and the star itself very small. It seems safe to assume that we have seen a red giant shedding its outer layer and revealing the white dwarf in its interior. It is likely that the star has been giving off matter for some time, but only now, after the white star with its high surface temperature has broken through, has the surrounding gas been made luminous. Thus, in planetary nebulae we are probably witnessing the birth of a white dwarf.

Stars do not always rid themselves of their hydrogen-rich outer layers so harmlessly. They can also shed them with explosive force.

Hartwig's Star in the Andromeda Nebula

Some advances in astronomy, those sparked by specific observations, can be dated with precision. One such momentous event took place on the night of August 31, 1885, when a 34-year-old German astronomer, Ernst Hartwig, at the Tartu (Estonia) Observatory, trained his telescope on the Andromeda nebula (see plate I). Hartwig knew no more about the structure of this spiral nebula than any of his contemporaries. That knowledge did not come for thirty-five years. When Hartwig located the nebula in his telescope, he noticed a star so bright as to be almost visible to the naked eye near the core, the brightest area of the nebula. That star had not been there the previous day.

He knew that stars occasionally flared up and then died down again, a phenomenon to which we shall return. But the remarkable thing here was that apparently this was a star within the Andromeda nebula. The finding became sensational in 1920 with the discovery that spiral nebulae, or galaxies, as they are known today, are accumulations of hundreds of billions of stars so far out in space that all but the most powerful telescopes show them simply as a nebulous veil of light rather than as individual stars. The Andromeda galaxy is so far away that it takes light 2 million years to travel down from it to us here on Earth. In other words the event Hartwig witnessed that night had taken place 2 million years earlier. After flaring up Hartwig's star shone 10 billion times brighter than the sun, which explains why despite the enormous distance it was almost visible to the naked eye. Hartwig had witnessed a flare-up of unimaginable dimension, far more powerful than the flashes of light called *nova phenomena* (with which we shall deal in chapter 9) occasionally seen in stars. Hartwig had observed in the Andromeda nebula what we now call a *supernova*.

Soon afterward, Hartwig left Tartu to take on a new assignment, the stewardship of an observatory bequeathed to the city of Bamberg, Germany, by a wealthy resident named Karl Remeis. It was a position Hartwig occupied until the 1920s.

I would like to ask the reader's indulgence for a brief personal digression. In 1954 Wolfgang Strohmeier became the director of the Bamberg Observatory, and I his assistant. When we went through the files we found two letters addressed to Hartwig dating from World War I. One, a deeply touching document, was from a young soldier with whom Hartwig apparently had been corresponding. The young man, whose most ardent wish was to become an astronomer, was in a hospital, almost blinded by an explosion and fearful that he would not regain his eyesight. His name was Hans Kienle (1895–1975); he later became the director of the Göttingen Observatory and the mentor of many famous astronomers, among them Ludwig Biermann, Otto Heckmann, Martin Schwarzschild, and Heinrich Siedentopf.

The second letter was written by a young man in Sonneberg, Thuringia. He, too, wanted to become an astronomer, but his father had made him leave school and join him in his business. But during the war his father's business had closed, and the young man felt free to offer his services to Hartwig. He was willing to work without pay if only Hartwig would take him into the observatory. Hartwig did.

Later this amateur astronomer began formal studies. His name was Cuno Hoffmeister (1892–1968), and he was to become the builder of Sonneberg Observatory. It was his observation of a comet in 1942 that furnished Biermann with the key to the discovery of solar wind. Among the thousands of variable stars discovered by Hoffmeister, two created a stir. One, the star BL Lacertae, is the first of a class of objects, floating far out between the galaxies, that is still not fully understood. The other, to which we shall come back, has become one of the loveliest objects of X-ray astronomy. Cuno Hoffmeister alas did not live to see these two of his variable stars achieve their fame.

Let us return to Hartwig's supernova. If a supernova can flare up in the Andromeda galaxy, then it is not far-fetched to assume that the same can happen in our own galaxy. Has a supernova ever exploded in the Milky Way? It is difficult to distinguish between a supernova and a normal, relatively harmless nova, because a nova that flares up close to us can seem far brighter than a distant supernova. But today we know that at least two supernovae have appeared in our Milky Way in modern times. In 1572 Tycho de Brahe observed a bright star in the constellation Cassiopeia, and in 1604 Johannes Kepler described a very bright star that had appeared in the constellation Ophiuchus and after a while disappeared again.* Both these sightings were supernovae comparable to the one seen by Hartwig in the Andromeda nebula. The supernova is a star that flares up with explosive force and hurls vast quantities of matter into space. There are many places in our Milky Way where gas masses fly apart at great velocities. We suspect that they were all caused by supernovae exploding a long time ago and that remnants of the explosion cloud are still visible. The best known of these is found in the constellation Taurus.

*It is one of those inequities of history that of all modern astronomers Tycho de Brahe and Johannes Kepler were the only ones fortunate enough to see and describe a supernova. Both of them had already gained renown; Kepler's three laws of planetary motion are still being taught in school today. As luck would have it both were handed a supernova on a platter, so to speak, while generations of less renowned astronomers have vainly continued to look for spectacular sightings that might bring them immortality.

The Crab Nebula and the Sino-Japanese Supernova

There is a small nebula in Taurus that unlike the Andromeda nebula is composed of a diffuse gas mass instead of individual stars—the so-called *Crab nebula* (see plate V). These gas filaments fly apart at high velocities, at times attaining speeds of thousands of kilometers per second. Since the dimensions of the nebula are known, the time of the actual explosion can be calculated. It apparently took place around the year 1000. Do we have any documentation that would support such an occurrence in Taurus that year? As a matter of fact Chinese and Japanese sources tell of a bright star that appeared in 1054 at the site of the Crab nebula. The star shone so brightly that for two weeks it was visible even in daylight. It was a supernova. Apparently, no mention of this is found in European archives. Whenever I get my hands on a history textbook, I look up the year 1054. I have therefore learned a lot about that year. I learned about people who died in 1054 of whom I had never heard that they ever were born. But I couldn't find anything about a bright star that had appeared in the sky. It seems incomprehensible that so impressive a manifestation should have gone unnoticed. Perhaps at the time there was no widespread interest in stars and their behavior, or perhaps Europe was undergoing a two-week spell of bad weather.*

In the explosion of a supernova an entire star apparently blows apart and hurls its matter, or at least a major portion of it, into space. Does that mean that the star is gone forever, or is some portion of it preserved? The answer to this question was found in 1968, and we shall come back to it in the following chapter. But before going on let us see what happens to the stellar matter blown or hurled into space.

---

*I have since learned that in 1054 Ibn Butlan, a doctor in Constantinople, held a celestial phenomenon responsible for an epidemic in which about 1,500 people died. The site of this strange manifestation was at the approximate site of the Crab supernova explosion. Apparently, something was sighted in Europe after all.

The Fate of the Matter Leaving the Star

Interstellar space in our galaxy is not a void. It is filled with clouds of gas and dust which, as we shall see in chapter 12, give birth to new stars. Some of the gas may have been there from the beginning of time; later, after stars formed out of that gas and in turn gave off matter, the interstellar medium mixed with the gases flowing out from the stars. In the stellar winds of developed stars, dust particles are formed by condensation. The star R Coronae Borealis, for example, emits sooty clouds that dim its light. Out in space gas atoms settling on the dust particles form a solid coat on them. The dust particles continue to grow until they are destroyed, some because they evaporate when they approach a hot star, some because they are subjected to the cosmic radiation of high-energy particles, and some because they collide with each other. The absorption of stellar matter alters the chemical composition of interstellar matter, infusing it with heavier elements that had built up in the stars. Consequently, the stars themselves largely determine the nature of interstellar matter, the matter out of which new stars are born.

In supernova explosions this infusion is particularly strong, as we shall see in chapter 11, since the matter being hurled into space is very highly developed. The velocities of supernova explosions are so great that the particles soon fill up the space of the galaxy. They are the particles of cosmic radiation found both in space and on our earth.

It was not until 1968 that we learned that in addition to the expanding luminous gas cloud and cosmic radiation produced by a supernova explosion another body also remains.

8

Pulsars Do Not Pulsate

An article in the February 1968 issue of *Nature* was startling enough to be picked up by the daily press throughout the world. In it a team working in Cambridge, England, under Anthony Hewish announced that it had picked up radio signals from space.

A New Radio Telescope Is Installed in Cambridge

Radio astronomy made great strides after World War II. Cosmic gas, particularly interstellar matter, emits and absorbs radiation in the range of radio waves. Like light, radio waves can penetrate the earth's atmosphere, opening up new possibilities of receiving information from the cosmos. Cosmic radio emission gives us information about the nature of our galactic interstellar matter and enables us to receive and investigate the radio emission of the gas clouds in other galaxies. Some stellar systems, the so-called *radio galaxies,* radiate with particularly high intensity.

Incoming radio waves are influenced by matter that emanates from the sun and is floating into interplanetary space as the solar wind alluded to in the preceding chapter. This wind is responsible for fluctuations in the radiation we receive from distant sources, just as movement in the earth's atmosphere causes the stars to flicker.

In the 1960s a new radio telescope was installed in Cambridge to study the fluctuations caused by interplanetary matter. More than two thousand aerials were installed in an area equivalent to fifty-seven tennis courts to study the solar-induced fluctuations of radio sources. The receiver was adjusted to record any change, however rapid, in radio emission. The old radio telescope was not able to do this. The new telescope was practically made to order for finding the rapidly changing pulsar signals—a discovery that pushed the original purpose, the examination of solar-wind–induced fluctuations, to the sidelines.

The large antenna field cannot be moved; it daily registers radiation by scanning the strips of sky that pass by its line of sight. Work on the installation was completed, and the telescope put into operation, in July 1967. Its receiver wavelength was at around 3.7 meters, and the intensity of incoming radiation was recorded round the clock. Seven zones of the sky were recorded on 210 meters of tape each week. The researchers were looking for uniformly emitting radio sources whose radiation, observed via the solar winds, scintillated. The telescopic observation and the tedious evaluation of the tapes were the responsibility of Jocelyn Bell, a doctoral candidate. She was looking for rapid fluctuations in the radiating celestial objects that passed by the sight line of the radio telescope with the movement of the earth.

Jocelyn Bell's Report

Nine years after the event Jocelyn Bell, who had since become Mrs. Bell-Burnell, reminisced in an after-dinner speech about her work with Hewish at Cambridge. She told of the endless tapes she waded through. After going through the first 30 meters, she was able to detect radio sources scintillating in the solar wind and differentiate between them and man-made radio interference:

Six or eight weeks after starting the survey I became aware that on occasions there was a bit of "scruff" on the records, which did not look exactly like a scintillating source, and yet did not look exactly like man-made interference either. Furthermore I realized that this scruff had been seen before on the same part of the records—from the same patch of sky.*

She wanted to pursue this lead, but other work interfered. By the end of October 1967 she was finally able to resume her investigation. She hoped to be able to do a better time study of the signal but could not find anything. Then at the end of November it reappeared.

As the chart flowed under the pen I could see that the signal was a series of pulses, and my suspicion that they were equally spaced was confirmed as soon as I got the chart off the recorder. They were 1 1/3 seconds apart. I contacted Tony Hewish who was teaching in an undergraduate laboratory in Cambridge, and his first reaction was that they must be man-made. This was a very sensible response in the circumstances, but due to a truly remarkable depth of ignorance I did not see why they could not be from a star. However he was interested enough to come out to the observatory at transit-time the next day and fortunately (because pulsars rarely perform to order) the pulses appeared again.

The pulse apparently was not of terrestrial origin, for the signals recurred whenever the same patch of sky passed by the telescope (see figure 8.1). Moreover, the pulses looked as though they were man-made. Did they perhaps come from another civilization? But they did not seem to be coming from a planet moving around another star.*

Just before Christmas I went to see Tony Hewish about something and walked into a high-level conference about how to present these results. We did not really believe that we had picked up signals from another civilization, but obviously the idea had crossed our minds and we had no proof that it was an entirely natural radio emission. It is an interesting problem—if one thinks one may have detected life elsewhere in the universe how does one announce the results responsibly? Who does one tell first? We did not solve the problem that afternoon, and I went home that evening very cross—here was I trying to get a Ph.D. out of a new technique, and some

*Jocelyn Bell-Burnell, "Petit Four," *Proceedings of the Eighth Texas Symposium* (New York: The New York Academy of Sciences, 1977), p. 685.

*If that were the case the interval between two pulses in the rhythm of the orbiting period of the planets would decrease and increase because of the varying distance of the radio source and the variable transit time of light. The effect is illustrated in another context in figure 10.5.

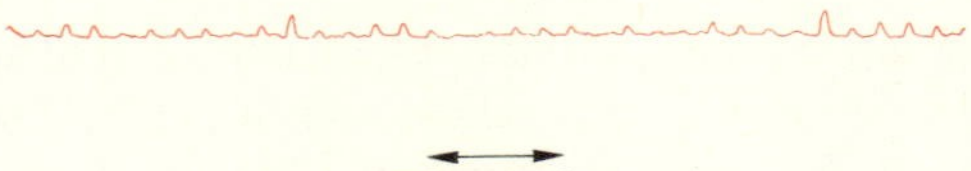

5 Seconds

Figure 8.1 The signals of one of the first discovered pulsars. Even though the individual pulses are not uniform, they nonetheless follow each other in precise, even intervals.

silly lot of little green men had to choose my aerial and my frequency to communicate with us. However, fortified by some supper I returned to the lab that evening to do some more chart analysis. Shortly before the lab closed for the night I was analyzing a recording of a completely different part of the sky, and in amongst a strong, heavily modulated signal from the well-known radio sources Cassiopea A I thought I saw some scruff. I rapidly checked through previous recordings of that part of the sky, and on occasions there was scruff there. I had to get out of the lab before it locked for the night, knowing that the scruff would transit in the early hours of the morning.

So a few hours later I went out to the observatory. It was very cold, and something in our telescope-receiver system suffered drastic loss of gain in cold weather. Of course this was how it was! But by flicking switches, swearing at it, breathing on it I got it to work properly for 5 minutes—the right 5 minutes on the right beam setting. This scruff too then showed itself to be a series of pulses, this time 1.2 seconds apart. I left the recording on Tony's desk and went off, much happier, for Christmas. It was very unlikely that two lots of little green men would both choose the same, improbable frequency, and the same time, to try signalling to the same planet Earth.

Soon afterward Jocelyn Bell discovered two more *pulsars*. At the end of January 1968 the Cambridge team submitted the manuscript containing the first public announcement of pulsars to *Nature*. Here is Jocelyn Bell's account:

A few days before the paper was published Tony Hewish gave a seminar in Cambridge to announce the results. Every astronomer in Cambridge, so it seemed, came to that seminar, and their interest and excitement gave me a first appreciation of the revolution we had started. Professor Hoyle was there and I remember his comments at the end. He started by saying that this was the first he had heard of these stars, and therefore he had not thought about it a lot, but that he thought these must be supernova remnants rather than white dwarfs.

At one point the article alluded to the fact that the Cambridge team had toyed with the idea that these signals might have been sent out by another civilization; naturally, the popular press pounced on this:

When they discovered a woman was involved they descended even faster. I had my photograph taken standing on a bank, sitting on a bank, standing on a bank examining bogus records, sitting on a bank examining bogus records: one of them even had me running down the bank waving my arms in the air—Look happy dear, you've just made a Discovery! (Archimedes doesn't know what he missed!) Meanwhile the journalists were asking relevant questions like was I taller than or not quite as tall as Princess Margaret (we have quaint units of measurement in Britain) and how many boyfriends did I have at a time?

Pulsars Are Small

The most startling discovery about the pulsars was how rapidly their emissions fluctuated. The brightness of variable stars with the briefest cycles can change within an hour or even faster. For instance, a white dwarf in the binary system of the 1934 nova in Hercules, which will be discussed more thoroughly in chapter 9, rhythmically changes in brightness in seventy-second intervals, a speed record that went unchallenged until the pulsars came along. With the growing refinement of time analyses in pulsar research, the pulses showed a fine structure, and it became apparent that within a pulse the radio intensity of the source can vary by ten-thousands of a second (see figure 8.2).

The rapidity of these variations opens up the possibility of learning more about the object that is sending out the pulse. To illustrate let us imagine a sphere so far removed from the observer that to the naked eye, or even through a telescope, it appears as a mere dot (see figure 8.3). That sphere may flare up for brief periods. What does the remote observer see? The flash travels with the speed of light. Since the distances between the various parts of the sphere and the

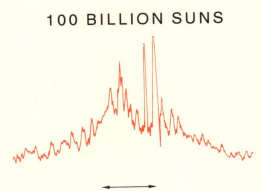

5 Milliseconds

Figure 8.2 A single pulse registered with high resolution in time. The pulsar signal reveals a complicated, detailed structure. (Adapted from H. D. Craft, Jr., J. M. Comella, and F. D. Drake)

observer differ in length, the emission, though simultaneous, reaches the eye at different time intervals. If we were able to see the sphere as an extended object it would appear as a disk. We first would receive the signal from the nearest point on the sphere, that is from the center of the disk seen by us. Subsequently a bright patch would develop in the middle, which would then change into a bright ring around the center of the disk. The ring's radius would increase until the ring reached the rim of the disk and disappeared. Figure 8.3

Figure 8.3 To the distant observer (right) a light pulse (top left) coming from a spherical surface appears smudged (bottom right) because the travel times of the light from various parts of the sphere differ.

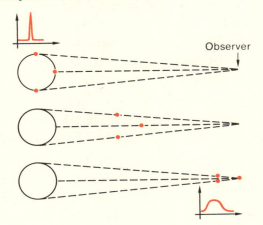

shows the first and the last light signals on their way to the observer. However, since we had assumed that the sphere is so far out in space that it appears as a point we cannot see the source of the signals in detail. But we would observe a variation in the time of arrival of the signals from that point in the sky. To the eye of the observer the brief flash appears smudged. It seems like a longer pulse. Its length is equivalent to the time it takes the light to traverse the radius of the sphere. But all light variations of the sphere, not only pulses, are smudged during this cycle, since all emissions, whether signaling an increase or decrease in brightness, have to travel distances of varying length. The smudging appears even if the body emitting the radiation is not spherical.

Thus, for example, if an emitting source shows fluctuations in intensity in the range of one ten-thousandth of a second, we must assume that the source cannot be much bigger than the distance the light travels in that time. That would be 30 kilometers. Were the source bigger the fluctuations would be smudged for longer periods. Within a single pulse the intensity varies within ten-thousandths of a second, as can be seen in the steeply sloping indentations in the recording curve in figure 8.2. Since the radio emission moves with the speed of light, we must conclude that the object sending forth the pulse cannot measure more than a few hundred kilometers in diameter, which is insignificant compared with the dimensions we are accustomed to in space. The diameters of white dwarfs can run to tens of thousands of kilometers; the earth's measures 13,000 kilometers. The pulsars are small cosmic bodies sending forth high-intensity radio signals.

After their discovery word about other newly discovered pulsars began to pour in from all over the world. Today we know of more than three hundred pulsars. Their cycles range from hundredths of a second to 4.3 seconds. Even though the form of the individual pulse may vary, the length of the cycle seemed to remain constant. Even when an individual pulse cannot be detected, the subsequent pulses will follow the established rhythm.

Since their discovery through refined time analyses, the individual pulses have been analyzed still more closely. Their structure was found to be even more refined than figure 8.2 indicates; the highest recorded speed of light fluctuation is 0.8 millionth of a second. The area sending forth the signals therefore measures 250 meters in diameter at most.

In the very years that pulsars were discovered, it was also found that their cycles lengthen. The pulsars slow down with time, but the intervals between pulses increase only slightly. On the average it takes 10 million years for a pulsar cycle to double in length.

Can Pulsars Be Seen?

What kind of objects are we talking about? Are they near the solar system or as far away as other galaxies? Obviously, they can be found among the stars of our own Milky Way system. We already know that the luminous ribbon of the Milky Way that we see in the night sky is made up of the many stars populating the disk of our galaxy. Within that ribbon the stars appear particularly dense in the area we perceive when we look toward the center of the galactic disk. If we plot the pulsars on a map of the heavens, we find that, like the stars of our galaxy, most of them are located in the ribbon of the Milky Way (see figure 8.4).

Their distribution in space thus parallels that of the stars: they are found among them. Presumably, then, some of the pulses have been traveling for thousands of years before being received by radio telescopes. But that would mean that the pulsars are incredibly powerful, considering that we are still able to receive their signals despite the vast distances involved. Yet the energy is supposed to originate in a very cramped space, perhaps no more than 250 meters in diameter. Immediately after the first pulsar was discovered and its location in the celestial sphere more or less precisely established, the presumed location was scanned by telescopes. A star in the area staked out by the radio astronomers proved to be completely normal. It ap-

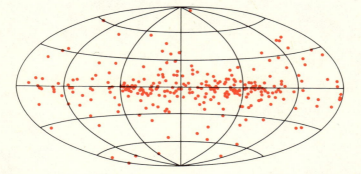

Figure 8.4 The distribution of more than three hundred pulsars in the sky. The grid has been chosen to show the entire celestial globe on an oval surface of the grid. The Milky Way stretches along the horizontal center line, and its center coincides with the center of the grid. Most of the pulsars are close to the Milky Way. (Adapted from Andrew G. Lyne)

parently had nothing to do with the radio emission coming from that direction. The pulsar itself remained invisible.

In the fall of 1968 pulsar signals with a three-hundredths of a second cycle were discovered coming from the direction of the Crab nebula. Pulsar emission was coming from the explosion cloud of the Sino-Japanese supernova of 1054! Do the stellar objects in the Crab nebula (see plate V and figure 8.5) have anything to do with the pulsar? Is one of them perhaps the pulsar?

How can one tell by looking at a star whether it is sending out pulses in the radio range? Does the light emitted by it in the optical range pulsate as well? In the case of so weak an object, the eye could never detect whether an emission is continuous or pulsating. Photography still offers no clues, for pictures collect the light of a star, whether steady or pulsating, at one spot on the film.

To determine whether or not the visible emission of a star pulsates demands special techniques. In principle one could put a TV camera behind a telescope and bring the optical signal onto two TV screens (see figure 8.6). We already know the pulsation cycle of the radio pulses. Thus, we could transmit the picture to Screen A for half that cycle and to Screen B for the other half. If the visible emission of an object varies in rhythm with the cycle of the radio pulse, the pulse

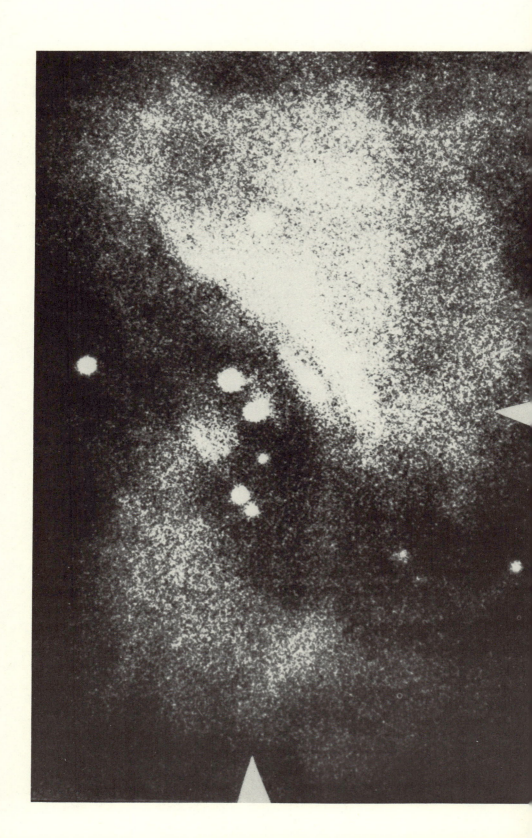

Pulsars Do Not Pulsate

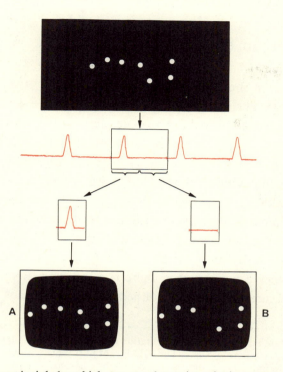

Figure 8.6 The principle by which one can determine whether a star sends out its light evenly or in pulses. On top, a view of the group of stars as seen without special devices. Below it, the periodic light flashes of one of the stars. The picture is televised to two screens, A and B, in the rhythm of its cycle. Screen A always receives the pulse, and Screen B receives the picture when the star does not send out a pulse. By comparing the pictures (bottom) it is possible to identify the star that sends out its light in pulses. (The method employed is here simulated with the help of a star in Ursa Major, which in fact is not a pulsar but sends out its light evenly.)

would always be flashed on Screen A, while Screen B would remain blank during the nonpulsating interval in which the object does not give off any visible emission. Those light sources that are not pulsating in rhythm with the cycle of the changing pulsar would appear equally bright on both screens. Thus, all one has to do is compare the pictures of the two TV screens to see whether any of the stars emits its light in the pulsating rhythm of the cycle of the radio pulse.

Figure 8.5 The central region of the Crab nebula. The area shown here in enlargement appears in the center of plate V. (Photographed by the primary focus of the Shane reflector of the Lick Observatory by J. D. Scargle)

137

The Pulsar in the Crab Nebula Becomes Visible

The pulsar in the Crab nebula was discovered by the method described above. The mechanism used operated on the same principle, except that instead of making a single picture of the sky, every star in the region being scanned was tested individually. The light of the star being analyzed was not distributed among a number of TV screens but, in rhythm with the Crab pulsar's cycle, was transmitted to a number of electronic radiation counters. The scheme of this type of measurement is shown in figure 8.7. If a star sends out its light uniformly, all counters receive the same number of photons. If a star emits flashes in rhythm with the cycle of Crab pulsar, only the counters that receive signals when there is a pulse are involved during the cyclical intervals, while the others remain idle. Thus, if the light

Figure 8.7 A similar schematic representation of a device to help determine whether a star sends out its light in pulses. The light is sent in rhythm with the cycle (known from the radio pulses) to counters that count the photons. The four boxes on the bottom represent four counters into which photons (red dots) are being "thrown." Every cycle is divided into four intervals. Photons received by the measuring device during the first interval are "thrown" into Box 1, those received during the second interval into Box 2, and so forth. After a cycle is completed the instrument starts anew at Box 1. In our example the third counter receives a great many photons whereas the others go almost empty-handed.

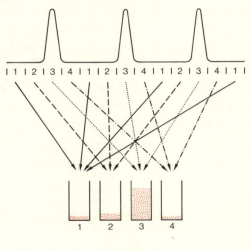

138

of the pulsar is distributed to the different counters over numerous pulsation cycles, those receiving light from the pulse will always register higher values, while those that receive light only from the night sky, which is never completely dark, register only low figures. This is known as a pulse *building up* in the electronic counter.

In November 1968 two young astronomers, William John Cocke and Michael Disney, asked to spend some nights at the 90-centimeter mirror telescope of the Steward Observatory at Tuscon, Arizona. Neither had any experience in observational astronomy, and they wanted to familiarize themselves with the telescope. While they were still debating about what to observe during their stint, the December issue of *Science* reported the discovery of the Crab pulsar. They decided to use their alloted time to look for the Crab pulsar in visible light. The observatory possessed the requisite electronic equipment. It had been built by Donald Taylor with an entirely different objective in mind, and since he held claim to this valuable electronic dowry he became a member of the observation team. From a technical point of view everything was in tip-top shape. Nonetheless, the project did not look too promising. No one had ever succeeded in identifying a visible star as a pulsar. Still, at least Cocke and Disney would learn how to work with a telescope and Taylor's electronic equipment would be given a chance to be tested.

In early January the measuring apparatus was set up at Kitt Peak; on the 11th everything was ready, and the telescope was trained on the Crab nebula for the first time. Each star was measured over five thousand pulse cycles, and during that time the light signals were directed to a number of electronic counters in rhythm with the radio cycle. None of the counters built up a pulse for any of the stars being tracked. Taylor went back to Tuscon on the 12th, and Cocke and Disney remained on the mountaintop, assisted by Robert W. McCallister, who kept an eye on the electronics. The weather worsened, and still there were no results. The following two nights of observation—the last two granted them for this project—had to be called off because of bad weather; the undertaking seemed doomed. But occasionally coincidence plays a major role. William G. Tifft, the observer who had claim to the telescope beginning January 15, became ill and offered to make the nights of the 15th and 16th available to the luckless investigators, so they tried once more. Here is Disney's account of what happened next:

The day of the fifteenth was cloudy but the sky showed signs of clearing toward evening, and observations were started on time (8 P.M.). As before Taylor was in Tucson, Cocke and Disney alternated on telescope and electronics, whilst McCallister operated the tape recorder and the computer. The first run, at 8 P.M., was on sky. During the second, centered on a star, a distinctive pulse appeared on the screen within 30 seconds and continued to build up. A secondary pulse some 180° out of phase, was also noticed, rather broader and not so high as the first. McCallister remained calm, but Cocke and Disney alternated between hysterical excitement and grim despair. Was this the pulsar or was it some electronic artifact? Working so close to 30 cycles, as we were, such artifacts were a distinct possibility.

In silence, Disney checked the telescope steering, Cocke the timing electronics and McCallister reset the computer for a repeat run. Within 30 seconds the pulse reappeared once again, with identical shape, to be greeted by a positive blast of Anglo Saxon profanity. Doubt crept in once more however. It was the next runs however which positively convinced the observers they were seeing a pulsar.

At this point (8:30 P.M.) Disney rang Taylor to give him the news. At first, Taylor was understandably sceptical, but as he heard all the results conviction began to grow. He suggested one or two circuit changes which might eliminate the possibility that faulty electronics were responsible. It wasn't until the next night however, when he saw the pulse grow for himself that he became convinced.

At 10:10 P.M. Cocke and Disney rang their wives who, by chance, were spending the evening together. It was only with difficulty that they were restrained from ascending the mountain there and then. Judging from the oohs and ahs there can be no doubt who were the two most excited people in Arizona that night.*

Others now came along to confirm the discovery. Figure 8.8 shows two pictures based on the principle demonstrated in figure 8.6. The pulsar missing from the photo on the right is the lower of the two stars in the center of figure 8.5. It is identified by markings on the right and the lower edge. These markings will help to locate the pulsar in the picture of the Crab nebula in Plate V.

*Personal Communication by Dr. Disney.

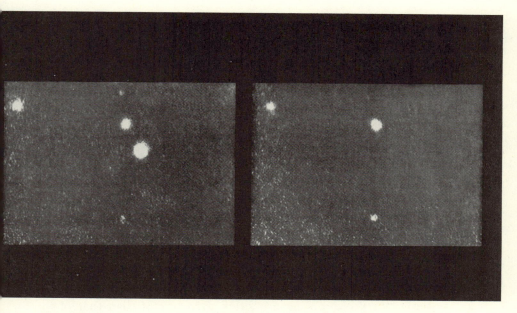

Figure 8.8 Two pictures obtained via the method employing two TV screens (see figure 8.6), which indicate that one of the stars in figure 8.5 is sending out its light in the rhythm of the Crab pulsar. In figure 8.5 two white markers, right and bottom, point to the star. By comparing figure 8.5 and Plate V, the Crab pulsar can easily be found in the color photograph; it looks no different than any of the other stars in the picture. (Photograph: Lick Observatory)

What Are Pulsars?

After the discovery of the Crab pulsar, it became evident that pulsars were somehow related to supernova explosions. Apparently, the remnants of a star after it explodes as a supernova send out pulsar signals. This assumption was supported by the discovery that pulsar signals were coming from another part of the sky in which the presence of gas masses pointed to a supernova some time in the past. This explosion in the constellation Vela must have taken place long before the Crab nebula supernova, for the gas masses hurled into space no longer appear in the sky as a single compact spot but as separate fibers in a vast space, as gas filaments. The 0.09-second cycle

141

of this pulsar is longer than that of the Crab pulsar. It is the third most rapid pulsar known. From the very beginning scientists looked for the object in the region of visible light. Not until 1977 did their search show results. The letter to the editors of *Nature* dated February 9, telling of the successful identification of the Vela pulsar with a star, was signed by twelve scientists. For eight years not only these twelve scientists working in England and Australia, but a considerable number of others, using the best available equipment, had been searching for the star sending out signals in the rhythm of the Vela pulsar. The success of the twelve put an end to a worldwide, laborious quest. Incidentally, Michael Disney of Crab pulsar fame was one of the twelve signatories.

As of this writing not a trace of any of the other pulsars has been found in visible light, which leads us to this conclusion: whatever pulsars may be, they originate in a star's explosion as a supernova. Initially, the pulsation cycle is brief, even briefer than that of the Crab pulsar. The pulsar sends out both light flashes and radio pulses. As time goes on the rhythm of the pulses slows down. In fewer than a thousand years the cycle lengthens to that of the Crab pulsar, and after still more time to that of the Vela pulsar. With the lengthening of the pulsing cycle, the pulsar's visible light diminishes. Later, when its cycle has grown to seconds and more, its optical flashes disappear, but it still makes its presence known by radio waves. That is why the only pulsars we have been able to see are two with very short cycles. They are among the most recent pulsars; the remnants of the explosion cloud are still visible. The oldest pulsars long ago lost their luminosity in visible light.

But what are pulsars? What remains when the life of a star comes to an end in a gigantic explosion? We already know that the space in which pulsar emission originates must be very small. What processes take place so quickly in so small a space and repeat with such regularity as to explain the pulsar phenomenon? Are they perhaps stars like the Delta Cephei stars that expand and contract? If so they must be very dense, for only then could their fluctuating cycles be that brief. We know that the cycles of Delta Cephei stars span days.

We, however, are looking for objects that can oscillate within hundredths of seconds. Even the densest stars, white dwarfs, cannot oscillate that rapidly, which gives rise to the question whether there perhaps exist stars of densities beyond any known to us.

The first speculations along these lines were offered long before there was even an inkling of pulsars by two astronomers working in Pasadena at what was then the biggest telescope in the world. One was the German-born Walter Baade, one of the foremost observing astronomers of our time, and the other, the highly imaginative though somewhat argumentative Swiss Fritz Zwicky. In 1934 they hypothesized the existence of high-density stars composed of practically nothing but neutrons. In 1939 the physicists Robert J. Oppenheimer and George M. Volkoff published a paper on neutron stars in the *Physical Review*. And long before astrophysicists began to work seriously on neutron stars, one of the paper's co-authors, Oppenheimer, was to gain worldwide fame as one of the architects of the atomic bomb.

Oppenheimer and Volkoff showed that matter in which all electrons and protons have fused into neutrons can form stellarlike gas masses held together by the force of their own gravity. These *neutron stars* can be calculated theoretically if the properties of the neutron matter are known. The stellar models of neutron stars show that they would be extremely dense. Their mass could be compressed into a sphere measuring 30 kilometers in diameter, and a cubic centimeter could contain billions of tons of neutron matter (see figure 8.9). If one were able to make neutron stars oscillate, their oscillation periods would be much shorter than those of pulsars. Thus, we can conclude that pulsars are not neutron stars.

Figure 8.9 A comparative-scale representation of the sun, a white dwarf, the earth, and a neutron star. Only the rim of the sun is visible at the top of the picture.

We are back where we started from. In the search for dense, stellarlike objects with the requisite rapid oscillations, we found that the white dwarfs were too slow and the hypothetical neutron stars too fast.

Thomas Gold Explains the Pulsars

To his friends and colleagues he is Tommy. He was born in Austria and emigrated to England in 1938, just before Hitler marched into Austria. He received his advanced training in England and prior to his move to the United States worked with another Austrian émigré, Hermann Bondi, and with Fred Hoyle. At the time the discovery of pulsars created a stir in the scientific community, he was teaching at Cornell University. The professional journals rushed into print with premature analyses and explanations, mostly efforts to salvage the pulsation hypothesis; Tommy Gold followed his own counsel.

The rotations of celestial bodies are among the most constant cyclical phenomenon in the heavens. The sun rotates about its own axis in twenty-seven days; other stars rotate much more rapidly. Are the regular cyclical patterns of pulsars perhaps connected with a rotational process? If so, the pulsar would have to rotate about its axis once per second, and in the case of the Crab pulsar 30 times. But a star does not have the option of rotating at whatever speed it wants, for if it did, centrifugal force would shatter it. Only stars with powerful surface gravitational force are able to rotate about their axes very rapidly—a white dwarf rotates about once per second at most. If it were to rotate in rhythm with the Crab pulsar cycle, its centrifugal force would shatter it. Only still denser stars can rotate more rapidly.

That brings us back to the neutron stars. Now the question arises whether the rotation of a neutron star is perhaps the built-in pace setter of the pulsars. A neutron star would then be rotating about its axis in fractions of a second, which is entirely possible given the power of its gravitational force. It could rotate even more rapidly, if pulsars are rotating neutron stars and the rotation is the pace setter. This was what Tommy Gold was suggesting.

Astrophysicists today consider Tommy Gold's hypothesis that pulsars are rotating neutron stars the most persuasive. Moreover, the gradually lengthening cycles of the pulsars would indicate that the rotation of neutron stars slows down over time. That seems to make sense, for the energy emitted by the pulsars, both in the radio range and in visible light, might well be fed by the rotational energy of the neutron star. The radio emission would thus gradually put a brake on the rotation of the neutron star, but the braking mechanisms are still more powerful. It is estimated that the rotational energy being released by the slowing down of the Crab nebula pulsar is strong enough to fuel the emission of both the pulsar and the nebula as a whole. This phenomenon also helps us solve another puzzling question.

While the light of ordinary gas nebulae, such as the planetary nebula in plate IV or the Orion nebula in plate VI, is emitted by atoms, the light of the Crab nebula is of quite different origin. There we find electrons that were accelerated to high velocities during the supernova explosion; they move almost at the speed of light. In the magnetic fields of the nebula, they are forced into circular orbits and emit their energy in the form of light. There always was the nagging question why these electrons should have been moving at such a rapid rate since 1054, why radiation had not slowed them down. If it had, their light emission would have diminished, and with it the brightness of the Crab nebula. They obviously receive energy from somewhere, and now we have discovered the source. If Tommy Gold is right, the Crab nebula contains a rotating neutron star which, perhaps through its magnetic field, supplies its gaseous neighborhood with energy. The neutron star sweeps through the nebula like an eggbeater, seeing to it that the electrons maintain their velocity and the Crab nebula its brightness. The neutron star contains enough rotational energy to last for thousands of years to come.

True, we have found a mechanism to explain the cyclical regularity of the pulsars, but we still lack an explanation of how the radio

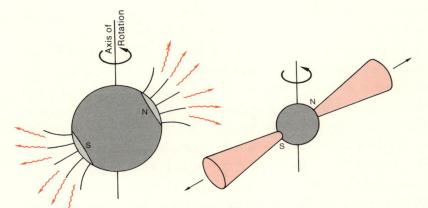

Figure 8.10 A possible model of the origin of pulsar signals. At the two magnetic poles (N and S) of a rotating neutron star, electrons fly along the magnetic-field lines into space with almost the speed of light. While in the vicinity of the star they radiate in flight direction. The photons emitted are shown at the left as red, wavy arrows. The radiation emitted by the neutron star thus goes out from the magnetic poles into space in the form of two radiating cones (right), which revolve along with the rotation of the neutron star, scanning the sky like the beams of a lighthouse. An observer receives radiation only if hit by one of the two cones and will see the neutron star flare up in rhythm with its rotation.

emission originates. Since we are not dealing with a simple wave but a pulse from which we receive no energy for protracted periods and then a great deal spasmodically, we have to visualize the star as sending out light in a particular direction, fixed with respect to the star and thus sweeping through space with the rotation of this star. We at regular intervals are picked out by the beam, like a ship by the moving beam of a lighthouse.

In all likelihood the neutron star possesses a magnetic field somewhat like the earth's, only far more powerful. We shall come back to this in chapter 10 when we discuss X-ray stars. Let us assume that the magnetic field of the neutron star is not aligned with its rotational axis. Neither is the earth's. When the neutron star rotates, its magnetic field moves along with it. (For a visual representation see figure 8.10.) On the surface of the rotating, magnetic neutron star, the neutrons are converted into electrons and protons, and strong electrical forces push the charged particles away from the star. These particles fly into space along the lines of the magnetic field. Their energy is enough to let the Crab nebula continue to glow a thousand years after its birth. Since charged particles cannot easily move at a right angle to the lines of the magnetic field, they generally leave the neutron star at its magnetic poles, and from there they fly outward along the curved lines of the magnetic field. Figure 8.10 gives us a schematic view of this. The electrons, as the lightest particles leaving the star, will attain the highest speeds, probably in the range of the speed of light. Flying so rapidly along curved paths they emit energy. This energy does not radiate out equally in all directions but is concentrated strongly in the flight direction of the electrons. As a result the emission from the star is in the direction of the lines of the magnetic force coming from the star's surface. It radiates outward in two conic areas. Since the magnetic field is involved in the star's rotation, the two emission cones also rotate. A distant observer receives the emission only when one of the two cones passes over him. The neutron star appears to him to flare up in rhythm with its rotation. In this visualization—which many astrophysicists believe to be fairly accurate—we are hit by the emission coming from the direction of the polar magnetic fields as though by the rotating beam of a lighthouse.

Unanswered Questions

In the spring of 1969 two observatories independently monitoring the arriving signal of a pulsar discovered that the pulsar interrupted the slow, regular growth of its period and shortened the interval between successive pulses (see figure 8.11). Yet the slowing down continued at the same rate. Having begun to accept the idea that we may be dealing with a rotating neutron star that is braked and slowed down by its interaction with the surrounding medium, we are now faced with the question of why it would suddenly accelerate even for a very short period, so that the change appears to be abrupt.

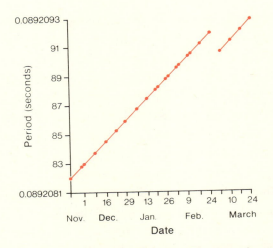

Figure 8.11 Cyclical glitch in the period of a pulsar. First the pulsar period increases slowly, then abruptly becomes shorter (top right), and then slowly increases again (P. E. Reichley and G. S. Downs)

The change is apparently spasmodic. Atomic physicists, who are more accustomed than astrophysicists to dealing with neutrons, believe that solid crusts, or clods, have formed on the surface of neutron stars which occasionally shatter during the cooling off of a neutron star left over from a supernova explosion. When the star subsequently contracts its rotational velocity increases. Is this perhaps the explanation for the many instances of abruptly shortened cycles that have since been discovered? Major changes in the earth's crust could change its rotational period and with it the length of the day. Are we witnessing a similar process in the pulsars? Are their cyclical irregularities a consequence of neutron starquakes?

In 1975 Russell A. Hulse and Joseph H. Taylor discovered a new pulsar. With a cycle of fifty-nine milliseconds it was the second most rapid after the Crab pulsar, but that is not its most important attribute. Its pulsing sequence is variable; three times a day the pulses come in shorter intervals, and in between in longer ones. This phenomenon is shown schematically in another context in figure 10.4, since it plays a significant role in the X-ray sources discussed in that section. That section furthermore points out that this phenomenon occurs when a source sending out a pulse circles around another celestial body, causing the distance the signals have to travel to reach us to vary—at times it is longer, at others shorter (see also figure 10.5). The Hulse-Taylor pulsar therefore revolves around another star. But no one has yet been able to see the star. That is not surprising, for the orbit of the pulsar is so narrow that there is hardly any room for the unidentified body. It cannot be a normal main-sequence star, let alone a giant or supergiant for they would be much too large. Does the binary pulsar perhaps revolve around a white dwarf or a neutron star? If the latter, is the neutron star not also a pulsar? We now know of three binary-pulsar systems. In the case of the other two the second star may possibly be a main-sequence star that survived the supernova explosion its companion underwent before becoming a pulsar.

The reader will recall that after the discovery of the first pulsar by Hewish's team, little green men were not considered the likely producers of the signals because the pulses do not change their intervals periodically (see footnote, p.129). That holds true for more than three hundred pulsars, but in the case of the three variable pulsars, the signals come from a body circling another star. How much greater would the excitement have been in 1968 had the Hewish team's first discovered pulsar been a binary pulsar.

In the last ten years great strides have been made in a new field of observational astronomy—gamma-ray astronomy. Gamma rays are electromagnetic waves of extremely short wavelengths. The wavelengths of the photons of this radiation are shorter than even shortwave X rays. Their emission is highly energetic; each gamma-ray photon contains about a million times more energy than a photon of visible light. Like X rays gamma rays do not enter our atmosphere from the outside; we first found out about cosmic gamma rays when we were able to look into space from rockets and satellites. One of the most startling facts learned from gamma-ray astronomy was that gamma rays are sent out by a number of pulsars. Because of their high energy gamma rays may be the most important aspect of the pulsar phenomenon, while radio rays, which brought pulsars to our attention, may be nothing more than a by-product, like the loud bang of an explosion. The rhythm of the gamma rays given off by a pulsar is the same as the rhythm of the radio pulses given off. But the gamma pulses do not coincide with the radio pulses! No explanation has as yet been found for the gamma-ray phenomenon of pulsars.

The pulsars pose another problem for astronomers. Based on the many pulsars already known it is estimated that our galaxy may contain as many as a million active pulsars. We have also been tracking remote galaxies for decades to find out how many supernovae on average explode in them in a hundred years. For once we know that, we can estimate the number of neutron stars that have recently originated in our own Milky Way. It now appears as if there are more pulsars in space than could possibly have formed through supernova explosions. Does that perhaps mean that pulsars can also originate via other means? Are some of them perhaps the result of more peaceful, less spectacular events than stellar explosions?*

In a recent interview Jocelyn Bell-Burnell told of a radio astronomer who before the discovery of the first pulsar had been observing a place in the region of Orion that we now know to possess a pulsar. The needle of his electronic apparatus oscillated rhythmically. He kicked it, and the oscillations stopped. That kick cost him dearly, for in 1974 Tony Hewish received the Nobel Prize for the discovery and for the investigation of pulsars by his group. That discovery was indeed remarkable, only the name given it was wrong, for pulsars do not pulsate. At the time they were named they were still believed to be somewhat like Delta Cephei stars, which expand and contract. Only later did we learn that they were rotating neutron stars. But things being what they are, they are now stuck with their name. Yet are we sure that Tommy Gold's hypothesis is correct? Do neutron stars actually exist? Astrophysicists continued to harbor doubts until X-ray stars were discovered. More of that in chapter 10.

*In November 1982 astronomers all over the world became excited when they learned that five radio astronomers using the Arecibo radio telescope in Puerto Rico had found a new pulsar that has broken the world record held until then by the Crab pulsar. During each second the new pulsar emits 642 single pulses; this means that a neutron star is rotating more than 600 times around its axis during each second. Indeed the gravity on the surface of a neutron star is sufficiently high that the centrifugal force cannot disrupt it even at that high a speed.

9

When Stars Steal Mass from Other Stars

Binary systems have proved to be fruitful objects of astrophysical research. We can learn more from them than from single stars. That holds true both for X-ray stars, the subject of the next chapter, as well as for ordinary stars that have joined together to form a binary system. For a time, however, it looked as though the binaries were proving our ideas about the evolution of stars to be fallacious. Some binary specialists asserted that stars evolve very differently than we had been led to believe by the computer simulations done during the 1950s and 1960s.

What gave rise to these doubts was a certain type of binary star that first attracted the attention of astronomers in 1667, when the Bolognese astronomer Gemiani Montanari observed that the second brightest star in Perseus occasionally appeared dimmer than usual.

Algol, the Demon's Head

Ptolemy called the star that Perseus, after whom the constellation is named, holds in his hand the Head of Medusa. The ancient Jews called it the Demon's Head; the Arabs, Ras al ghul, or "unquiet spirit"; and our name for it, Algol, is derived from the Arabic. Monta-

100 BILLION SUNS

nari recognized that it was a variable star. More than a hundred years later, on November 12, 1782, the 18-year-old John Goodricke in England noticed that the star was only about one-sixth as bright as it had been the previous night; the next night it had regained its normal brightness. On December 28 the variation in brightness recurred. At 5:30 P.M. Algol was faint, yet three and a half hours later it again turned bright. Goodricke continued his observations and soon found the key to its waxing and waning brightness. Algol normally is bright, but every sixty-nine hours, for a period of three and a half hours, it grows dimmer until it is only a little more than one-sixth its normal brightness; and over the next three and a half hours, it again grows brighter until it reaches its normal brightness.

Goodricke's explanation is as valid today as it was then. In an essay in the London Royal Society's *Philosophical Transactions,* the gifted young scientist, who incidentally could neither hear nor speak, wrote: "If it were not perhaps too early to hazard even a conjecture on the cause of this variation, I should imagine it could hardly be accounted for otherwise than either by the interposition of a large body revolving around Algol, or some kind of motion of its own, whereby part of its body, covered with spots or such like matter, is periodically turned towards the earth."*

A hundred years were to pass before his discovery gained credibility. Today we know that the first of the two explanations he had given was right. Every sixty-nine hours a companion star in the course of its orbit moves in front of Algol and partially obscures it.

This phenomenon can be seen by anyone who knows where in the sky to look for Algol. Because the star is usually bright it does not attract much attention. But periodically it appears as faint as its normally far weaker neighbor Rho Persei.

We now know of many variable stars that like Algol are periodically eclipsed by a companion; Zeta Aurigae, mentioned in the opening chapter, is one of them. Eclipsing variables are such cramped systems and so far away that the two stars cannot be seen separately even through the most powerful telescopes. But the course of the

Philosophical Transactions 73 (1783): 474.

Plate I. The distance between the Milky Way system, to which all the stars seen here belong, and the Andromeda galaxy, which looks like a nebular blotch, is about 2 million light-years. Only the most powerful telescope will reveal that the nebula is composed of countless stars. Many galaxies show the spiral structure we find here. Viewed from the Andromeda galaxy our Milky Way would look just as the Andromeda galaxy does here. (© California Institute of Technology, 1959)

Plate II. A view of the spiral nebula M51 in the Hunting Dogs constellation. The bright spirals are areas in which interstellar gas is made luminous by bright blue stars. The light from this constellation travels 12 million years before it reaches us. (Photograph: U.S. Naval Observatory, Washington, D.C.)

Plate III. The constellation Pleiades (Seven Sisters). Its brightest stars make the neighboring gas masses luminous. In the picture the luminous nebulae overpower the light of the stars within. (The four rays emanating from the bright stars in the picture and the ring result from deficiencies in the photographic equipment, not from any properties of the stars.) In addition to the bright, visible stars, more than a hundred other stars are part of this system; all of them move through space at the same speed. Presumably, they all originated at the same time. (© California Institute of Technology, 1961)

Plate IV. The planetary nebula NGC 7293. The luminous red matter was discharged by the weak star in the center of the ring. The central star resembles a white dwarf. The other stars are in space, either in front of or behind the spherical nebula, and have nothing to do with it. (© California Institute of Technology, 1961)

Plate V. The Crab nebula is the remnant of a supernova observed in the year 1054. Because of the long time it took its light to travel down to us, the star there exploded even before the Sumerians settled in Babylonia (ca. 4000 B.C.). (© California Institute of Technology, 1959)

Plate VI. The luminous gas nebula in Orion. In an area of about 15 light-years in diameter, the interstellar gas is highly concentrated; a cubic centimeter of that gas can contain as many as 10,000 hydrogen atoms. Even though this is a very high density for interstellar matter, the gas in the Orion nebula is still much more rarified than the best vacuum that can be produced on Earth. The luminous gas is estimated to consist of 700 solar masses. The gas in the nebula is made luminous by luminous blue stars. The Orion nebula contains stars that undoubtedly are a mere one million years of age. In the nebula are regions of higher density that justify the assumption that stars are still forming there. The light we are seeing from it now emanated approximately 1,500 years ago. (Photograph: U.S. Naval Observatory, Washington, D.C.)

Plate I

Plate II

Plate III

Plate IV

Plate V

Plate VI

eclipse tells us something about the pair, and what we learned about the so-called *Algol stars* seemed to contradict accepted beliefs about the evolution of stars.

Complicated Forces in Binary Systems

If two stars closely revolve around each other then matter is subjected not only to gravitational force, which pulls everything toward the center of the one star, but also to the gravitational pull of the companion. Moreover, centrifugal force can also become a significant factor in the movement of the stars around each other.

The gravitational pull in the neighborhood of a star therefore changes in a very complicated manner when a second star is involved. The French mathematician Edouard Roche at Montpellier more than a century ago developed a reductive method used by astrophysicists to this day to describe the forces in close binary systems.

A single star simply pulls all neighboring matter toward its center by gravitational force. But in a binary system the second star also exerts pull toward its center. On the straight line connecting the two centers the gravitational forces of the two bodies pull in opposite direction. They cancel each other partially or completely (see figure 9.1). Let us call the two bodies Star 1 and Star 2. Since the pull of a mass diminishes over distance, the force of Star 1 will predominate in the immediate vicinity of Star 1, but the pull of Star 2 will predominate in its own vicinity. It is thus possible to define a "permitted" volume, the volume around each of the two stars within whose confines gas brought to each would fall completely to that star. In this permitted volume, generally referred to as the *Roche volume,* the pull of the star always dominates. The cross section of the surface of the two maximally permitted volumes results in the broken line shown in figure 9.1. But we should keep in mind that centrifugal forces also influence the gas as it participates in the orbital movement of the stars around each other. Matter outside the two Roche volumes of

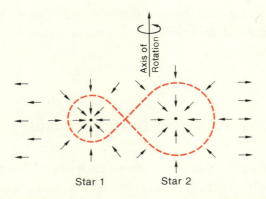

Figure 9.1 The forces in a close binary. The two stars are represented by black dots. The arrows indicate the direction in which the force affects a gas atom. Near each of the two stars its own gravity predominates; the arrows point to either Star 1 or Star 2. On the line connecting the two stars there is a point at which the gravitational forces of the two cancel each other. Since the two stars rotate around each other (the axis of the orbital movement and the direction are indicated on top of the drawing), centrifugal force, which seeks to throw matter outward, predominates at a greater distance from the axis (left and right). Each star has only one maximal volume at its disposal. When it expands and exceeds the area described by the red broken loop, parts of its outer layer flow over to the other star. The maximal permitted volume of a star in a binary system is also referred to as its *Roche volume*.

figure 9.1 can be hurled out of the system through centrifugal force, or it can fall on either of the stars. Matter within the Roche volume of a star *must* rain down on that star. The extent of the permitted volume depends on the mass and distance of the two stars and can be easily calculated for known binaries.

In observing close binaries we frequently come across systems in which the two stars are much smaller than their Roche volumes [see figure 9.2(a)]. The dominant force on the surface of each of the two stars is the gravitational pull toward their respective centers. One might say that neither of the two is aware of its companion. It is therefore not surprising that these close binaries—called *detached systems*—are not unlike single stars. Most of them are ordinary main-sequence stars living off hydrogen fusion that have used up only a small portion of their fuel.

Yet we also know binary systems in which only one star can be observed with certainty within its permitted volume, while the other

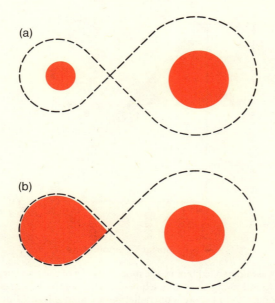

Figure 9.2 (a) A detached binary. Both stars are clearly located within the maximal permitted volume outlined by the broken curved line. (b) A semidetached binary. The left star completely fills out its Roche volume.

---

one just about fills its volume. These are the so-called *semidetached systems* [see figure 9.2(b)], of which the Algol system is one. These are the systems that pose a problem.

The Paradoxes of Algol and Sirius

In semidetached systems the more massive of the two stars is smaller than its Roche volume. It is a normal main-sequence star, unlike its less massive companion, which just about occupies its permitted volume and which in the H-R diagram is found to the right of the main sequence, having clearly moved in the direction of the red giants (see figure 9.3).

This fact would seem to overturn all our ideas about stellar evolution. After all, we found that the more massive stars evolve more

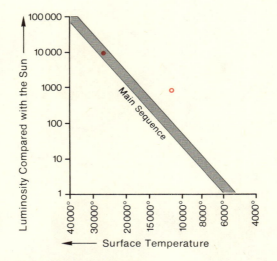

Figure 9.3 In a semidetached binary the more massive of the two components (filled circle) is still found along the main sequence, whereas the less massive one (open circle) is no longer found there, contradicting a commonly held notion about the evolution of stars that the more massive one is supposed to leave the main sequence first.

rapidly and use up their hydrogen sooner. Now we find two stars of identical age, and the less massive of the two shows symptoms of depletion before its more massive counterpart. There can be no doubt that the two stars originated at the same time, for a star cannot capture another star. Why, then, does the less massive one evolve more rapidly? Can it be that our basic ideas about stellar evolution are wrong after all? Moreover, it would appear that our ideas about stellar evolution run into trouble not only with Algol stars. Some detached binaries have exhibited traits that proved even more embarrassing.

Take Sirius, for example: we know that Sirius forms a binary system in conjunction with a white dwarf of only 0.98 solar mass. In connection with the computer calculations of the evolution of the sun, we found out that a star of less than 1 solar mass can turn into a white dwarf in about 10 billion years at the earliest. This means that the Sirius companion is considerably older than our sun. The primary star of Sirius, on the other hand, is composed of 2.3 solar masses and should therefore have evolved far more rapidly, yet it still shows all the properties of an undeveloped, hydrogen-burning star. Here, again, we have a system in which the more massive of the two stars is not yet showing signs of depletion while its less massive companion has reached a stage of advanced depletion.

Sirius is not an aberration by any means. We know of many binaries in which a less massive white dwarf is yoked to an undeveloped companion star of more mass.

Binary Systems in the Computer

Of course, no one wants to cast doubt on the basic theory of stellar evolution. After all, its findings corroborate observed phenomena, as with star clusters, for example. What, then, is responsible for the confusion in the evolution of a star that is part of a closed binary system and not of a cluster in which the stars are far apart? The only possible explanation is gravitational pull. Its main effect is not the deformation such close binaries are bound to suffer, for only the outermost layers deviate from the normal spherical shape of single stars, and these layers have practically no effect on the star's evolution. Rather, it is the inhibiting effect on the star's free growth.

Let us imagine a star in a close binary system that continues to expand for whatever reason, until it ultimately fills its permitted volume. Any further expansion would cause part of its surface layer to intrude into the Roche volume of its companion. At that point matter from the expanding star must flow to the companion. This is the new insight into the evolution of close binary stars: stellar mass can in time flow from one star to the other with almost sudden speed, for every star expands when energy-producing nuclear reactions deplete the hydrogen content of its central region. Thus, if we have a binary system that initially is clearly separated [see figure 9.2(a)], the more massive star will deplete its hydrogen sooner and be ready to turn into a red giant. However, it soon reaches its maximal volume.

If it then continues to expand, mass must flow to its companion. What happens at that point cannot easily be predicted.

Once again we call on the computer for help. Actually, everything proceeds pretty much as in the evolution of single stars. All we have to do is to make the computer understand that the space available to the star is finite. The computer must determine that available volume at each stage of the star's evolution and compare it with its volume. If the star is too big, the computer subtracts matter from its surface and calculates a model for the now shrunken star and then adds the subtracted mass to its companion. The transfer of mass from one star to the other changes the gravitational pull, the orbiting period, and thus the centrifugal force of the two stars. Therefore, the computer must now recalculate the permitted volume available to each and test whether, after the exchange of mass, each is content to stay within its Roche volume or whether mass continues to flow from one to the other. In this way the computer can simulate the evolution of stars when transfer of mass is involved, which has given us the needed tool for studying the story of binary systems, using different examples.

The first solution of the Algol paradox was offered by Donald Morton, a student of Schwarzschild at Princeton, in early 1960 in his doctoral dissertation. Subsequently, we learned how to follow troublesome phases of stellar evolution on computers; in 1965 Alfred Weigert and I began to work on the problem in Göttingen and calculated a series of binary systems. Two of the stories of their evolution follow.

The Story of the First Binary System— A Semidetached System Is Born

Our hypothetical pair began its career as two main-sequence stars of 9 and 5 solar masses, respectively, revolving about each other in a 1.5-day cycle at a distance of 13.2 solar radii. The more massive of the two evolved first, whereas the other evolved at an almost imperceptible rate. As the 9-solar-mass star used up more and more of its hydrogen, its outer layers slowly expanded. After 12.5 million years the hydrogen content in its center had shrunk to half its original amount, and its expansion continued apace, until it filled up its permitted volume. In the H-R diagram in figure 9.4, the evolutionary stage at which it now finds itself is designated *a*. If expansion continues even minimally, the process is unstoppable: mass must flow out to its companion.

The calculations indicate that a minor transfer of mass is not enough to decrease the star's volume. A catastrophe is in the making, and it lasts for sixty thousand years. During that time the star loses

Figure 9.4 The evolution of a close binary whose components are composed of 5 and 9 solar masses, respectively. In the more massive member the depletion of hydrogen becomes apparent first. It would—following the course of the red dotted line—actually like to turn into a red supergiant. However, it attains its maximal permitted volume before the hydrogen at its center is completely depleted. Going through a phase of rapid mass exchange, it moves along the broken curved line toward b, while the star gaining mass wanders upward along the main sequence. The originally more massive star, now the less massive of the two, completes the fusion of hydrogen at its center between b and c. By the time it reaches c it has shrunk to 3 solar masses, whereas its companion has grown to 11 solar masses. (The respective mass content—in solar-mass units—is indicated along the evolutionary track and the main sequence.)

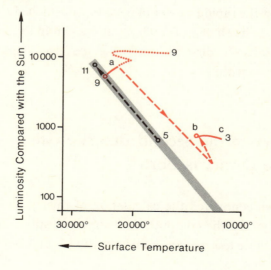

5.3 of its original 9 solar masses to its companion, which now is composed of 5 plus 5.3, or 10.3, solar masses. The companion has absorbed so much stellar gas that it has become the more massive of the two. The time involved in the role change of the two stars, the more and the less massive, is brief in stellar terms. The depleted star now stands at point *b* in the H-R diagram. Earlier, while it still was the more massive of the two, it had already consumed a substantial portion of its hydrogen. It is now an evolved star, and consequently, it is found to the right of the main sequence. Having settled down there, it embarks on a period of slow evolution during which it uses up the remaining hydrogen at its center. It expands slowly, and in the course of the next 10 million years continues to transfer mass to its companion.

The companion, having taken on mass, slowly begins to age. Still, it will remain on the main sequence for some millions of years longer. In that time our system shows the typical symptoms of Algol stars: the more massive star, barely evolved, lies on the main sequence, while its less massive companion, having already departed from the main sequence, fills up its permitted volume.

The exchange itself takes place in one two-hundredth the time of the evolutionary processes that precede and follow it. For that reason we can see only systems *prior to* (in the form of detached systems) or *following* the rapid transfer of mass (semidetached systems) in our Milky Way. The chance of catching such a pair in the act is marginal. Donald Morton's dissertation, presented five years earlier, had rightly foreseen this.

The Story of the Second Binary System—
A White Dwarf Is Born

When we conducted the calculation of this system Klaus Kohl, who later joined the computer industry, was still a member of our team. We chose less massive bodies, putting 1- and 2-solar-mass stars at an initial distance of 6.6 solar radii. The results are shown in the H-R diagram in figure 9.5 and in a scale model in figure 9.6.

Here again, the more massive star evolves first, and in the course of its evolution its radius continues to increase. But the distance chosen here is such that the star reaches its maximum permitted volume after all the hydrogen at its center has been converted into helium. After 570 million years the star enters a critical phase. Like its counterpart in the semidetached system, it undergoes a very rapid transfer of mass (lasting 5 million years) in which approximately 1 solar mass flows from it to its companion, followed by a continuing slower transfer (lasting 120 million years), at the end of which only 0.26 of the initial 2 solar masses remain. The star has lost almost its entire hydrogenous outer layer. What remains is the helium that had

---

Figure 9.5 The birth of a white dwarf. Let us assume that the more massive of the two stars—it consists of 2 solar masses—begins its course along the main sequence at a, and its companion—a 1-solar-mass object—at a'. The more massive star evolves first, reaching its Roche volume at b. While giving off mass to its companion, it moves along the course indicated by the broken line. At d the mass exchange comes to a halt, and the originally more massive star, which by now has been reduced to 0.26 solar mass—turns into a white dwarf on the way to e. As it gains in mass its companion moves along the main sequence toward d'. (See also figure 9.6.)

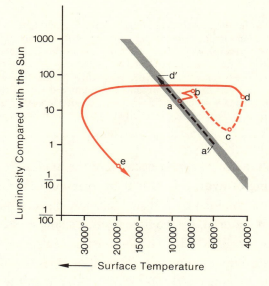

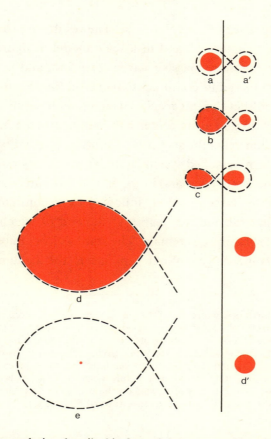

Figure 9.6 The evolution described in figure 9.5, represented here in uniform scale. The letters correspond to the relevant points in the H-R diagram of figure 9.5. The maximal (Roche) volumes available to each star are outlined by the black broken lines. In the course of mass exchange the distance between the two stars can change noticeably. When the two stars are further apart their permitted maximal volumes are also greater. The vertical line represents the rotational axis of the binary. If at the outset (top) two main-sequence stars revolve about each other, then we find at the end (bottom) a main-sequence star (right) and a tiny white dwarf (left).

formed at its innermost center because of the nuclear reactions during the burning of hydrogen. Our 0.26-solar-mass star consists of a helium core surrounded by a distended, low-density outer layer of hydrogen-rich material. By the time it stops losing mass, the star has turned into a red giant. The calculations make it possible for us to look into this giant star, a luxury observation does not afford us. Almost its entire volume of 10 solar radii is filled with the highly rarified gases of its hydrogenous outer layer; 99 percent of its mass consists of helium compressed into a small central sphere one-twentieth the diameter of the sun. It is the white dwarf in the red giant. But the outer layer of our star still is distended. By the time it stops losing mass it has exhausted its expansive force and gradually settles down onto the small, central helium sphere. In the process its radius contracts substantially, and even seen from the outside the star increasingly takes on the attributes of a white dwarf. It moves to the lower left of the H-R diagram, toward the region of the white dwarfs.

What, meanwhile, has happened to its companion? After all, during the loss of mass of the initially more massive star, the companion gained 2 minus 0.26, or a total of 1.74, solar masses. Once again the roles of primary and secondary star are reversed. After its increase in mass the now more massive star (2.74 solar masses) still has not had time to evolve significantly, whereas the other has already turned into a white dwarf. The calculations in fact offer proof that in a system of two stars of like age, a white dwarf can indeed exist together with a more massive main-sequence star, as seen in the Sirius system.

The apparent paradoxes and difficulties have been resolved. The observed behavior of binary stars give further proof that the basic theories about stellar evolution are on the whole valid.

The masses and the separation of many of the detached systems are such that with the onset of hydrogen depletion in the more massive member, the subsequent outflow of mass will follow the described pattern, and ultimately a white dwarf will be formed.

It is, however, by no means certain that the story of a binary system recounted here, ending with a white dwarf, describes what actually happened in the Sirius system. As we know, stellar winds or the formation of a planetary nebula can free single stars of their outer layer and turn them into white dwarfs. Perhaps the Sirius system never underwent a transfer of mass, perhaps the originally more massive star shed its outer layer without outside assistance, and perhaps only a fraction of that mass went over to its companion while the major portion escaped into space. Yet even if this is what happened, the paradox is solved, for in that event, given its great mass, the originally more massive star evolved more rapidly than its now more massive companion. In either case the now less massive star originally was the more massive one.

The transfer of mass between members of binary systems also plays a part in the nova phenomenon. These stars with their powerful light explosions had already been spotted in ancient times, but only recently, after 1945, was persuasive evidence presented that they most likely are close binaries.

The August 29, 1975, Nova in Cygnus

Anyone looking up into the sky on Friday, August 29, 1975, who had at least an inkling of the major constellations would have noticed something strange going on in the constellation Cygnus. A star was present that simply did not belong there. It was seen first in countries in the East, where darkness had already fallen and the stars were out. When dusk fell over the West, the new star obviously attracted the attention of observers there as well (see figure 9.7). Amateur astronomers rushed to their telescopes and the professionals to their observatories. Was it possible that an event expected since Kepler's day was actually taking place? Had a supernova finally exploded in our Milky Way? Were we witnessing the birth of a neutron star, like the one created by the Crab nebula supernova?

Figure 9.7 The explosion of the nova in Cygnus on August 29, 1975. The dots represent individual brightness observations at specific times.

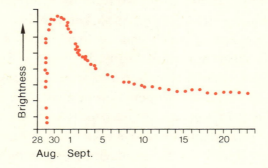

By now the star in Cygnus is nothing more than an insignificant, faint object, visible only through a telescope. As it turned out it was not the longed-for supernova but an ordinary nova.

This minor, harmless version of the supernova phenomenon was first noticed in 1917 when one learned from old photographs that two stars had flared up in the Andromeda nebula in 1909. They were 1,000 times weaker than the supernova discovered by Hartwig in that galaxy fourteen years earlier. We now know that its burst of energy corresponded to that of other flashing stars that periodically appear in our Milky Way. A particularly lovely example of one such outburst in the Milky Way occurred in 1901 in the constellation Perseus.

Novae, the name given these flashing stars, have nothing in common with supernovae. They are considerably weaker but occur far more frequently. In the Andromeda galaxy they can be seen to flare up anywhere from 20 to 30 times a year. In looking at old photographs of the sky, we found that stars used to stand where novae have flared up. A few years after it flares up, a nova again takes on the properties of the star it once was. A nova is simply a star that flares up and later returns to its original state.

In the neighborhood of a dimmed nova, it is not unusual to find a small, rapidly expanding nebular cloud obviously thrust out during the flare-up. But unlike the clouds of exploding supernovae, the nova cloud contains only a small amount of mass. The star did not explode; it lost only a little mass, probably somewhat less than a thousandth of its total.

The Nova of 1934

What kind of stars are they, these insignificant celestial objects that suddenly flare up ten-thousandfold, then over the next few months grow dimmer until after a few years they revert to their old, insignificant selves?

Typical of such a star is the nova that flared up in December 1934 to become the brightest star in the constellation Hercules. By April of the following year its brightness had dimmed perceptibly; it then became a little brighter but remained below the naked-eye level of visibility. Today it can be seen with a medium-range telescope.

What can this now faint object teach us? Probably its most important lesson is that on closer inspection the former nova turned out to be a binary system. That discovery was made by Merle F. Walker in 1954 at the Lick Observatory. The two stars revolve around each other in an orbiting cycle of four hours thirty-nine minutes. Because they eclipse each other in the course of their revolutions, we know still more about them. One of them is a white dwarf of 1 solar mass, the other is most probably a less massive ordinary main-sequence star. But there are still more surprises to come. The main-sequence star just fills its permitted volume, and from its surface matter flows out to the white dwarf. What we have here is a semidetached system similar to that of the Algol stars, with gas flowing out from one star to the other, but in this case the matter flows onto a white dwarf.

And we know still more. The matter does not settle down on the dwarf immediately. Since the entire system is revolving, centrifugal force inhibits the motion of the overflowing gaseous matter, which gathers initially in a disk that flows around the white dwarf. From that disk, matter slowly rains down on the dwarf (see figure 9.8). We cannot see the disk directly, but during the orbital cycle the main-sequence star slowly moves over the circular disk and covers it section by section. This partial eclipse is reflected in the diminution of the observed light of the system, to which the luminous circular disk also contributes. Not only has the structure and expanse of the ring been studied but we also know its temperature to be particularly high at the very spot at which the matter exuded by the main-sequence star meets with it. The ring has a hot spot originating at the place at which the gas flow from the main-sequence star is moderated, since a portion of its motion is transformed into heat. Furthermore, it appears that the white dwarf in the Hercules nova changes in brightness in a seventy-second cycle.

Every former nova studied more closely turned out to be a binary system in which a white dwarf was being showered with matter from a main-sequence star. There also exist novalike stars, the so-called

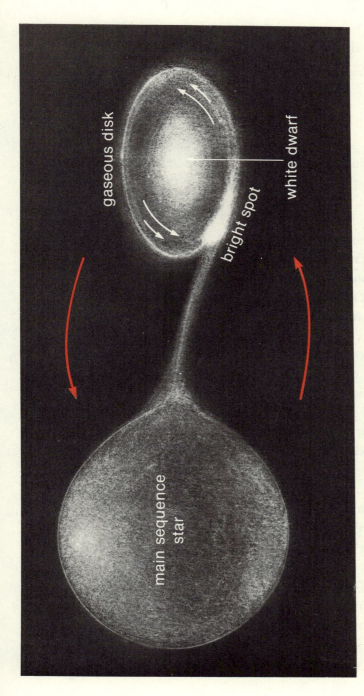

Figure 9.8 The two components of the binary in a nova orbit about each other as outlined by the red arrows. A main-sequence star fills its maximal permitted volume. Gas from its surface is given off to its white-dwarf companion. Initially, however, the matter rotates in a disk before settling on the white dwarf. The white, curved arrows represent the movement in this disk. At the place at which the gas flow from the main-sequence star collides with the rotating disk, a hot, bright spot is seen. (An artist's view, drawn by Hans Ritter)

dwarf novae. Their flare-ups are substantially weaker and recur in irregular intervals. They, too, are binaries of the type just described.

Nuclear Explosions in Binaries

What in a binary is responsible for the sudden release of a vast amount of energy, enough energy to make the stars shine 10,000 times brighter for brief periods?

The thinking that led to the answer to this question goes back to Martin Schwarzschild, to Robert Kraft (now at the Lick Observatory), and to calculations made by Pietro Giannone (presently at the Rome Observatory) and Alfred Weigert (currently the director of the Hamburg Observatory) at Göttingen in the 1960s. The theory has been further elaborated by Sumner Starrfield and his team at Arizona State University at Tempe.

The interior of a white dwarf is still hot enough to ignite hydrogen, but when that white dwarf formed in the heart of a red giant, all the hydrogen there had long since turned into helium, and the helium had possibly even converted into carbon. Consequently, the interior of a white dwarf does not contain hydrogen. But the gas that rains down on the white dwarf from its neighboring main-sequence star is rich in hydrogen. It falls on the relatively cool surface of the dwarf star, whose temperature is not high enough to ignite the hydrogen. Thus, a thick layer of hydrogenous matter forms on the surface and continues to grow. At the same time, the base of this layer heats up at the point at which it touches the original matter of the white dwarf. This process continues until the temperature at that site reaches about 10 million degrees. When it does the hydrogen ignites and the entire hydrogenous layer is thrust out into space with powerful explosive force. The computer calculations of Starrfield and his colleagues followed this hydrogen bomb on the surface of a white dwarf as far as possible, and it looks as though they found the explanation for the nova phenomenon.

When Stars Steal Mass from Other Stars

The fact that the flare-ups of some—perhaps even of all—novae recur lends support to that conclusion. Thus in 1946 a nova was found in the constellation Corona Borealis which had flared up once before, in 1866; there are others that flared up several times (see figure 9.9). These occurrences fit in with our picture. After the explosion, the main-sequence star, which itself has not been impaired in any way, continues to feed the white dwarf with hydrogenous matter. Once more a highly explosive layer forms on the surface of the dwarf, which is exploded again when the hydrogen ignites.

Whether or not the ex-nova of the 1975 flare-up in Cygnus was a binary is still not certain. And astrophysicists are beginning to consider the possibility that a hydrogenous layer can form on a single white dwarf when interstellar gaseous matter collects on its surface. But these reflections are probably premature. Perhaps the system must first calm down before a binary system of the sort seen in other ex-novae can become visible. However it is possible that even then we will still not know. For if we look down vertically on the orbital plane of the binary system we cannot identify the movement of the two stars around each other by their Doppler Effect (see appendix A), nor does it announce its duality by the mutual eclipsing of its components.

The close binary systems in which gas flows from one star to the other have introduced a number of new phenomena. The seeming paradoxes of Algol stars, the riddle of the age of binary systems such as Sirius, have been solved. But they are also responsible for possibly the most exciting celestial bodies known to us, X-ray binaries.

Figure 9.9 The explosions of the nova T Pyxidis recur in irregular intervals. They have been observed in the years 1890, 1902, 1920, 1944, and 1966.

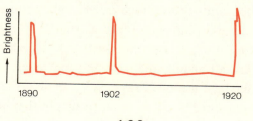

10

X-ray Stars

This chapter deals with stars that, unlike the sun, do not send out
visible energy. Their radiation lies in wavelengths beyond the percep-
tion of our senses, radiation of which we had no inkling until Wilhelm Conrad Röntgen discovered it accidentally in 1895. X rays, like
light and radio waves, belong to electromagnetic radiation but their
wavelengths are much shorter. Radio waves have wavelengths measuring meters or even kilometers. The short light wavelengths, measured in ten thousands of millimeters, are long compared to those
of X rays, which are a hundred times as short.

On first hearing, it might seem strange that X rays should originate
in space. We are all acquainted with the technical equipment involved in the medical use of X rays. Is it possible that these same
rays come to us from the cosmos? In principle they are produced

by the same process. In the medical apparatus a sudden brake is put on high-speed electrons to produce the radiation. In nature a gas is heated to millions of degrees, causing its electrons to move at high velocities. If such an electron comes near an atomic nucleus, its motion is suddenly altered by the electrical field of the nucleus, producing the same sort of radiation as is produced in the X-ray tube.

The gas shroud surrounding the sun, called the *corona,* has a temperature of about 2 million degrees. In it rapidly moving electrons are braked and then accelerated by collisions with atomic nuclei, and every time that happens roentgen radiation is born. The solar corona sends X rays into space, which satellites have been able to photograph. Thus, even so harmless a star as our sun has shown that X rays can originate in space. Only an insignificant part of the sun's energy is emitted in the form of X rays, however. X-ray stars, on the other hand, are tiny celestial objects that send out most of their radiation in the X-ray range. Their existence has been discovered only recently, and everything we have been able to find out about them since makes them exciting objects of study.

The Uhuru Story

Before the X-ray radiation coming from space can reach the earth's surface, it is absorbed by the uppermost layers of the atmosphere. That is why X-ray astronomy did not begin until it became possible to send balloon-borne remotely controlled telescopes to the top layers of the earth's atmosphere, or to launch them by rockets. The earliest X-ray experiments simply measured the radiation of the solar corona, but before long we directed our attention to other cosmic sources of X-ray emission. A new chapter in modern astrophysics was beginning.

Generally speaking, it is difficult to identify the wellspring of a major leap forward in science. The era of spectacular individual achievement is past, especially in the experimental sciences. Most innovative research is done by teams of scientists going to conference after conference, following up on the seeds of ideas, combining them with their own and coordinating the results with their colleagues'. When they finally co-author and publish a paper, all we see are the results; only rarely do we learn how they were arrived at.

The story of the discovery of X-ray stars easily fills a book. It was the subject of Richard F. Hirsh's doctoral dissertation at the University of Wisconsin. I will confine myself to just some of the milestones and to some of the scientists involved in the development of X-ray astronomy, and will consider one commerical enterprise as well.

In practically every airport passengers and their baggage are required to pass through a low-level X-ray security checkpoint. In North America most of these contraptions are manufactured by AS&E, American Science and Engineering, which was founded by Martin Annis in 1958, and is staffed primarily by scientists. In its early years AS&E worked on problems of nuclear energy in cooperation with MIT. The first X-ray satellite built was the work of AS&E. The key event in its building probably took place in 1959 when a young Italian exchange student met a famous compatriot in the United States.

Riccardo Giacconi had come to Indiana University at Bloomington from Milan in 1956 on a Fulbright fellowship. His special field and the subject of his doctoral dissertation of 1954 was the measurement of cosmic rays. From Bloomington he went to Princeton. Impressed by the high level of scientific research in the United States and by the funds being made available to space science in the wake of Sputnik, he decided to stay on in the United States. He made the acquaintance of Martin Annis, then the head of a 27-man research institute. In September 1959 Giacconi joined the staff of AS&E. A short time later Annis introduced him to Bruno Rossi, who had come to the United States before World War II. In Chicago Rossi had worked with the great Enrico Fermi on the development of the first nuclear reactor in Chicago. Now, in addition to his affiliation with MIT, Rossi headed an advisory panel at AS&E. Giacconi says that when he met Rossi for the first time, Rossi told him that in his opinion celestial X-ray emission was an important object of space research. Thus encouraged, Giacconi set about finding out all there was to know on the subject. What he found was that not much had been done. Herbert Friedman's studies of solar X-ray emission (carried out at the U.S. Naval Research Laboratory (NRL) in Washington, D.C.) were about all that had been done; no other cosmic source of X rays had been discovered.

Giacconi toyed with various potential X-ray receivers and in conjunction with other scientists studied other possible techniques for measuring cosmic X-ray emission. In 1960 NASA gave the go-ahead for the first X-ray telescope. Giacconi had already gathered about him a small team that had worked on space experiments at AS&E; by 1961 the team numbered seventy. By 1962 nineteen rockets and seven satellites had been launched carrying experiments of the team. One of these scientific loads was an X-ray receiver. It was found that the sun was not the only source of X rays; they also came from deep inside the Milky Way and even farther out in space. In July 1962 a pointlike source was found in the constellation Scorpio. The first X-ray star had been discovered. Here is the account that Giacconi has given to me: "Spurred on by our results, Friedman and the NRL scientists were able to confirm our findings as early as April 1963. In September 1963 I presented to NASA a plan for future research. In it, I described my concept for a new, slowly spinning satellite to be used for X-ray observations (Uhuru) and a 1.2 meter telescope for the future. Though Uhuru did not fly until 1970 and the Einstein-satellite not until 1978, the broad lines of my research approach were already clear. It remained for nature to make it all so worthwhile and exciting."

On December 12, 1970, NASA launched in Kenya a satellite built by Giacconi and his team. It was the seventh anniversary of Kenya's independence, and in honor of that occasion the satellite was named *Uhuru*, Swahili for *freedom*. Figure 10.1 shows a NASA representation of the space vessel Uhuru. During its sojourn in space Uhuru discovered more than one hundred pointlike sources in the sky. The

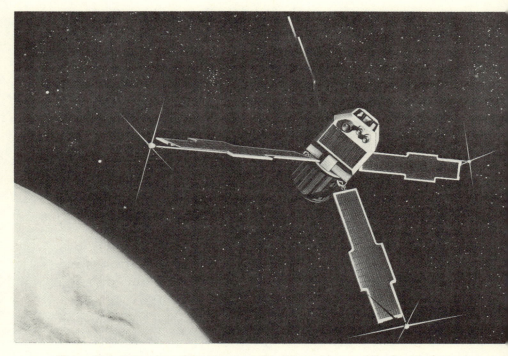

Figure 10.1 A drawing of the X-ray satellite Uhuru in space. The four "solar paddles" furnished the vehicle with solar energy. While the satellite rotated about its axis once every ten minutes, the X-ray receivers in the body scanned the sky in strips. The results of this exploration were radioed back to Earth. (Courtesy of NASA)

---

successful mission of this satellite brought Riccardo Giacconi and his team worldwide recognition, but it posed many problems for astrophysicists the world over. We are still a long way from understanding the objects Uhuru found, though we are learning more about them all the time.

The first question astronomers are likely to ask about newly discovered objects is whether they are nearby or far away. Determining precisely the distance to a celestial body is usually very difficult, but sometimes even approximations are useful. For example we would like to know whether or not these objects are located within our galaxy. As our research on pulsars has showed us, this can be determined by their distribution in the sky: does it follow the pattern of the stars of our galaxy? The result of this test is shown in figure 10.2,

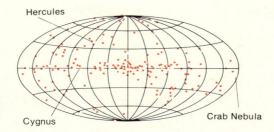

Figure 10.2 The distribution in the sky of the X-ray sources discovered by Uhuru. As in figure 8.4, the grid used here makes possible the representation of the sky in a flat oval. The Milky Way stretches along the horizontal center line. In the middle of the oval we are looking into the very center of the Milky Way disk. In the sky most of the X-ray sources are close to the Milky Way, primarily near the center. Some of the sources mentioned in the text are indicated here. (Adapted from E. M. Kellogg)

where the sources discovered by Uhuru are entered onto a grid of parallels and meridians in which the symmetrical plane of the Milky Way forms the horizontal center line. As we can readily see most of the X-ray sources are found in the vicinity of the Milky Way. Where stars are numerous X-ray sources are relatively abundant. But some sources can be found by looking out into space from the plane of the Milky Way, particularly in places where remote galaxies are more numerous.

I will confine myself to X-ray sources in the Milky Way. We know their approximate distance. On average they are as far out in space as most of the stars in the galaxy, that is to say, some thousands of light years away. The radiation reaching us makes it possible to estimate their actual radiative strength. In the X-ray range their strength is about 1,000 times as great as that of the sun in all wavelengths.

The X-Ray Star in Hercules

Let us begin by looking at a source in Hercules—Hercules X-1—discovered by Uhuru. The radiation received from it by the satellite was not uniform but came in X-ray flashes in 1.24-second intervals (see figure 10.3). The interval between two X-ray pulses

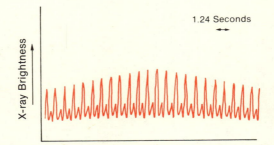

Figure 10.3 The X-ray flares of the source in Hercules discovered by Uhuru. (Adapted from H. Tananbaum et. al)

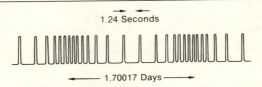

Figure 10.4. A schematic drawing of cyclical character of the intervals between the pulses of Hercules X-1 revealing a 1.7-day period. From this phenomenon we learned that the source is a binary.

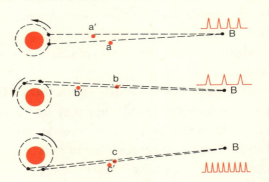

Figure 10.5 An imaginary X-ray source orbiting a star—here represented by a red disk—is sending out flashes in one-second intervals. An observer, B, at a great distance from the star, is measuring the intervals between flashes. (top) Two flashes, a and a', traveling down to the observer. When a' was being emitted the source stood at a different place in its orbital path than at the emission of a. In the model shown here the paths of both flashes are of equal duration. The observer perceives them in one-second intervals. (center) The distance traversed by the flashes b and b' emitted in one-second intervals differs. The distance traveled by b', which is sent out after b, is greater. The observer sees the two flashes in intervals of more than one second. (bottom) Flash c', sent out subsequent to c, travels a shorter distance; to the observer the flashes arrive in intervals of less than one second.

does not follow a rigid scheme, however. Periodically, it gradually grows shorter and then increases again, and this pattern repeats itself in a 1.70017-day cycle. Figure 10.4 gives a schematic view of this phenomenon. It appears that the X-ray source is first moving toward us and then away again, as it would if it were revolving around another celestial body.

Let us imagine an X-ray source with a one-day orbital cycle describing a circular orbit around a central star. The source itself might be sending out X-ray flashes at one-second intervals. In figure 10.5 we can see how the observer might then see the pulses in alternately longer and shorter intervals, precisely as in the X-ray source in Hercules. This leads to the conclusion that the source is moving around another star—in an orbital cycle of 1.70017 days.

By now the reader already knows what happens next: when two stars in a close binary system revolve around each other they can—from our point of view—occasionally eclipse each other. The result is a variable eclipsing star like Algol or Zeta Aurigae. If our X-ray source were revolving around a star, it could disappear behind its companion every 1.70017 days in the course of its orbit, at which point the X-ray signals would not be received.

That in fact turned out to be what was happening with Hercules X-1. Figure 10.6 shows us pictures of Uhuru observations made in

---

Figure 10.6 The pattern of the source Hercules X-1 during a protracted period of time. The dots represent the level at which the X-ray flashes were measured by Uhuru. Pairs of vertical lines at a five-hour distance from each other are plotted at intervals of 1.70017 days. They show the five-hour intervals in which no flashes are detected because the source stands behind the object it is orbiting. The pulses were not seen until January 9, and again not after January 21. This is related to the thirty-five-day cycle of the Hercules source mentioned in the text. (Adapted from Giacconi et al.)

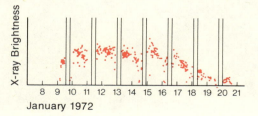

January 1972. Every 1.70017 days the X-ray pulses ceased for about five hours, the length of time the source was eclipsed by the other star.

But things become more complicated still. The X-ray source does not always radiate. It is "plugged in" for about twelve days, during which time its 1.24-second pulses can be measured, and is then interrupted by the five-hour eclipse. X-ray radiation then ceases altogether for twenty-three days, after which the whole business begins all over again.

The Hercules Source Becomes Visible

What is the nature of the site in Hercules at which the X-ray pulses originate? Uhuru was able to make only an approximate determination of its location. As figure 10.7 shows, many stars populate the area in question. Does one of these perhaps have attributes that merit

Figure 10.7 The region in the sky in which Uhuru discovered the Hercules source. Hoffmeister's insignificant variable star is indicated by the red arrow.

special study? The American astronomer William Liller was the first to call attention to a star that had been classified as variable in 1936.

At this juncture we once again meet Cuno Hoffmeister, the young man whom Hartwig had admitted to the Bamberg Observatory during World War I. In 1936 Hoffmeister had spotted a variable star in Hercules on celestial photographs. Hoffmeister had long since finished his university studies and had his own observatory, partly privately financed, and systematically roamed the heavens with his telescope in search of variable stars. He found thousands of them. The Hercules star did not appear to be in any way extraordinary. Hoffmeister did not find that the changes in brightness adhered to any overall pattern, that they might possibly be periodic. Later, he followed the star for a few nights, but it appeared not to vary in brightness any more. In 1936 Hoffmeister's object entered the catalogs as the variable star HZ Herculis, and no one paid any further attention to it. But in 1972, because of its proximity to the newly discovered X-ray source, interest in it rekindled. The X-ray source seemed to have an orbital cycle of 1.70017 days, and this gave rise to the question whether Hoffmeister's star perhaps followed the same pattern. In the summer of 1972 John and Neta Bahcall at the Tel Aviv Observatory established that the variability of Hoffmeister's star did in fact conform to this cycle. This supported the idea of a link between the visible star HZ Herculis and the X-ray source Hercules X-1. The visible star is fainter when the X-ray pulses disappear, that is, when the source stands behind it. It becomes brighter when, as seen from the earth, the source passes in front of it (see figure 10.8). This would explain the light variability. When the X-ray source is in front of the visible star, the side of the star facing us is heated by the intense X-ray emission and thus appears brighter. When the source is behind the star, it heats the side that is turned from us. Apart from this heat effect the star accompanying the X-ray source is an ordinary main-sequence star of about 2 solar masses.

How could so experienced an observer as Hoffmeister have failed on second observation to see his star as a variable object? Old celestial photographs show that at times the star did not vary in bright-

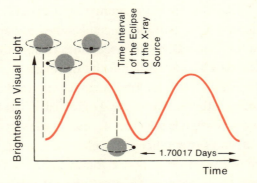

Figure 10.8 Hoffmeister's variable star HZ Herculis periodically changes in brightness (red curved line). Four sketches in the left segment of the figure (three above the curve and one below) indicate the respective position of the star (gray disk) and X-ray source (black dot) in relation to the observer. When the X-ray source—seen from the earth—stands in front of the star we see the side being heated by the source; it appears bright. When the source stands behind the star we see only the "normal," not the heated side, and the star appears faint.

ness for years. When that happened was it not being irradiated by the X-ray source? Does the X-ray source Hercules X-1 also stop radiating during this period? Since Uhuru's discovery of the source, HZ Herculis has varied in brightness in rhythm with the orbital cycle. But probably before too long it will again shine with constant light for years. Then we shall find out how the X-ray source behaves.

X-Ray Stars Are Small

Cygnus X-1, a source in Cygnus, behaves altogether differently. Instead of sending out regular pulses it shows irregular, very rapid variations in intensity. Moreover, its intensity also changes within months. A variable radio source stands in the same part of the sky, and its fluctuations are quite similar to those of the X-ray source: as the X-ray source changes in intensity, the radio source does, too, and when the radio source remains constant, so does the X-ray source. It is therefore safe to assume that we are dealing with one and the same object. In recent years radio astronomers have developed procedures that allow us to measure with great precision the position in the sky of a radio source. The exact position of the X-ray source could therefore be measured and its identity with a visible star unequivocally established. This star, too, is part of a binary system. Not that both stars can be seen—only one is visible—but the Doppler effect in its spectrum (see appendix A) indicates that together with a companion—in all probability the X-ray star—it revolves around the system's center of gravity in 5.6 days.

A few X-ray stars have temporarily appeared, only to disappear again. Centaurus X-4 was one that shone briefly. The source showed pulses in rhythmic 6.7-minute cycle; then after a few days died down again.

How do X-ray sources fit into the picture we have built up about happenings in space? They obviously are stellarlike objects. But how is it possible for a star to send out X rays? The surface temperatures of even the hottest stars known are far too low to give rise to strong X rays. The X-ray emission of the thin hot outer layer of some stars is, like that of the solar corona, far too weak.

X-ray pulses are very brief. In Hercules X-1 radiation rises to its maximum in less than a fourth of a second. The irregular fluctuations in Cygnus X-1 take place within hundredths of seconds.

Yet as we found with pulsars the rapidity of variations in brightness tell us something about the size of the radiative source. That holds true for the X-ray emission of the sources discovered by Uhuru as well as for light radiation and radio emission.

Because fluctuations in Cygnus X-1 vary within hundredths of seconds, the area emitting the X rays cannot be significantly larger than the distance light travels in that amount of time, that is, a distance of fewer than 10,000 kilometers, less than one-hundredth of the sun's radius. Consequently, these objects, which shine 1,000 times brighter than the sun, must be extremely small, a conclusion supported also by the rapid eclipsing of the Hercules source. That object disappears abruptly behind the star without any transition.

Considering the small size of X-ray sources, it seems plausible to

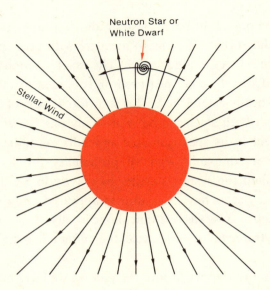

Figure 10.9 A possible cause of X-ray emission in binaries. A star (center) is sending out stellar wind into space. The black lines and arrows indicate the direction of the wind. Part of the gas mass ejected by the main star is captured by an orbiting neutron star or white dwarf and forced to fall on its surface at high speed. On impact the matter heats up enormously and X-ray emission results.

assume that they might be white dwarfs or neutron stars. If so the source of X-ray emission is not hard to figure out. As we know, it takes temperatures in the millions of degrees to produce X rays. Because of the force of gravity matter raining down on a white dwarf or possibly even on a neutron star hits the surface at great speed, which can easily generate temperatures in the millions of degrees on impact. This would be a likely explanation for the origin of X-ray emission. But where does that matter coming down at such high velocities originate? Is the explanation to be found in the fact that perhaps most or even all X-ray stars are part of a binary system? When a normal star and a white dwarf or neutron star revolve around each other, and when the normal star, such as the sun and many more, releases matter into space, a part of that outflow is sucked in by the gravitational pull of the companion star; this matter will rain down on the surface of the companion and in the process become hot enough for X-ray emission to occur (see figure 10.9).

The Story of an X-Ray Source

We can form an approximate picture of the history of an X-ray source. The story might go something like this: two stars of different mass have been revolving around each other for some time (see figure 10.10). The more massive of the two depletes its hydrogen supply first; it is ready to turn into a red giant. However, through the out-flow of matter into space or transfer of mass to its companion [figure 10.10(a)], it turns into a white dwarf [figure 10.10(b)]. We now have a binary pair composed of one main-sequence star and one white dwarf. Once the main-sequence star has also used up its hydrogen

---

Figure 10.10 Two possible stories of binaries that may explain the origin of X-ray sources. (left) Two main-sequence stars of different mass orbit each other closely. The more massive one is the first to show signs of depletion. It would in fact like to turn into a red giant. In the process its outer layers intrude into the gravitational field of its companion, which takes so much mass from it (a) that all that is left is its core, a white dwarf (b), as shown in figure 9.6. Now if the star to its right has turned into the more massive one, and in the course of its evolution produces stellar winds, part of the gas flowing out falls on the white dwarf, and X-ray emissions (red, wavy arrows) are a likely result (c), as shown in figure 10.9 (right) (d) Two stars of different mass are orbiting each other. The more massive one evolves more rapidly and explodes in a supernova. (e) The outer layer of the more massive of the two flies out into space; a neutron star remains and revolves with its companion, which is still a main-sequence star. (f) The main-sequence star of the binary evolves, and gas in the form of stellar wind flows into space. A portion of it falls on the neutron star, as shown in figure 10.9, and produces X-ray radiation (red, wavy arrows).

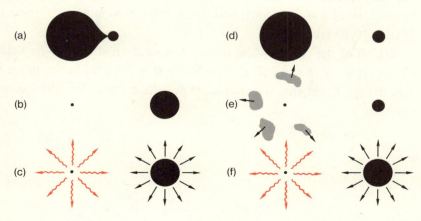

and turned into a red giant, the resulting expansion may cause it to exceed its permitted maximum volume, and its compact companion will then take mass away from it. Matter falls down on the compact object, and X-ray emission is produced. As little as a hundred millionth part of a solar mass raining down on the white dwarf in a year is enough to produce this effect. But it is also possible that a stellar wind flows out from the surface of the normal star and falls in part onto the white dwarf, thereby producing X-ray emission [figure 10.10(c)].

This scenario is reminiscent of the story of the Mira companion referred to earlier. The white dwarf revolving about Mira collects matter. Why, then, is it not an X-ray source? Perhaps because it is too far away from Mira to collect more than a small portion of Mira's gas flow—enough to make it shine in visible light but not enough to transform it into an X-ray star.

On the other hand it is also possible that white dwarfs play no role whatever in X-ray stars. One of the two stars of a binary system may conceivably have been shattered by a supernova explosion [figure 10.10(d, e)], leaving in its wake a neutron star like the Crab pulsar that now revolves around the comparatively unharmed surviving companion star. If this survivor now transfers matter to the neutron star, whether in the form of stellar wind or as a result of having overstepped the limits of its maximum volume during its evolution, the gas falls onto the neutron star's surface with even greater energy than in the case of a white dwarf, and the X-ray emission is still stronger [figure 10.10(f)].

Which of the two, a white dwarf or a neutron star, is to be held responsible for the emission from X-ray stars? Astrophysicists tend toward the belief that neutron stars are the answer. We shall soon see why.

Where Do Pulses Come From?

It seems plausible that X-ray emission can originate in cosmic sources, but we have not yet explained why it comes in the form of pulses. In the case of pulsars we believe that the rhythm of the pulse is caused by the rotational motion of the neutron star. Our compact object, like most celestial bodies, probably has a magnetic field, and like the earth its magnetic axis probably does not coincide with its rotational axis. Cosmic matter can move at an angle to the magnetic field line only with difficulty. The matter in the binary system will therefore fall on the compact object at its magnetic poles, as shown in figure 10.11. X-ray emission originates only where matter rains down, that is, in the vicinity of the magnetic poles. It can escape only sideways, for in the direction of the magnetic axes it is swallowed up by falling matter. When the compact object rotates the X-ray emission cannot be seen by a remote observer looking down onto one of the two magnetic poles. Between times it reappears again, as shown in figure 10.11.

Measuring the Magnetic Field of a Neutron Star

In our discussion of pulsars we assumed that magnetic fields are responsible for their radio pulses, and we again fall back on them with regard to X-ray stars. Where do the magnetic fields of neutron stars come from?

Magnetic fields are found almost everywhere in space. The sun has a large magnetic field similar to the earth's, except that it is about twice as powerful. The sunspot fields are more than 1,000 times more powerful. Other stars also give evidence of possessing magnetic fields.

Magnetic fields and cosmic matter adhere to each other. When a body grows denser so does its magnetic field, becoming more powerful in the process. If a white dwarf forms in the central part of a

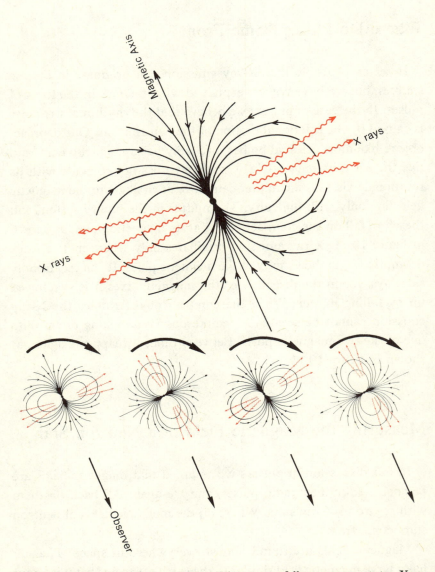

Figure 10.11 The origin of X-ray flashes. When matter falls on a compact star, X-ray radiation is produced on impact (upper segment of the figure). If the star's magnetic field resembles the earth's, that is, if its field lines are like the oval lines in the illustration, the gas can travel only along the field lines in the direction of the arrows to the two poles of the compact star, here represented as a small black sphere. The falling matter near the magnetic axis forms two "plugs" opaque to X-ray radiation. The radiation produced in this manner can escape only sideways into space (red, wavy arrows). When the entire object rotates it is possible that a remote observer will receive X-ray radiation only twice briefly within one orbital cycle. The imaginary, distant observer toward the bottom right will receive X-ray radiation only heavy black arrows. (For simplicity's sake we assumed the magnetic axis to be perpendicular to the rotational axis.)

star, the initially weak magnetic field in tandem with increasing density becomes 10,000 times more powerful. Strong magnetic fields have in fact been found in white dwarfs. When stellar matter contracts to the density of a neutron star, the magnetic fields grow more dense and can become 100 billion times more powerful. Scientists thus expected to find such powerful magnetic fields in neutron stars. And they were not disappointed.

On May 2, 1976, a balloon carrying scientific measuring instruments developed by the Max Planck Institute for Extraterrestrial Physics at Garching near Munich in conjunction with a team at the University of Tübingen was launched in Palestine, Texas. The team, under the direction of Joachim Trümper, had been working on X rays for some time and now wished to test a number of devices, including a new type able to detect higher-level X-ray emission than the detectors aboard Uhuru. The energy of X-ray photons is generally measured in kiloelectron volts (keV). The Uhuru receivers were able to "see" in the range of 2 to 10 keVs; the new receiver was able to detect X-ray photons of more than 30 keVs of energy. The Hercules X-1 source was the object of observation on that flight in the spring of 1976, aimed at investigating the radiative intensity at high levels of energy.

The more refined the technique, the less immediate the contact between the observer and his data. In 1936 Hoffmeister was still able simply to look through his telescope, estimate the brightness of HZ Herculis (the visible counterpart of Hercules X-1), and based on his notes decide there and then whether the star had or had not grown brighter since his previous observation. Now the results of measurements are fed into a computer by magnetic tapes, and computer programs have to be set up to read and interpret the information on the tapes. It is therefore not surprising that the results of the May flight did not become available until the fall. When they were read it was found that the radiation, which decreased in intensity at higher levels of energy, showed a peculiar notch at around 58 keV (see figure 10.12). This would most likely have been ignored if Trümper had not remembered something from a previous study of the radiation

187

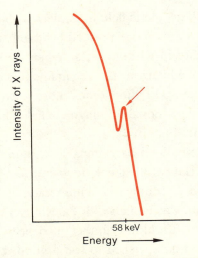

Figure 10.12 The X-ray emission of Hercules X-1 at high energy. Normally, one would expect that at higher energy the X-ray emission of the source would diminish. At 58 keV, however, a notchlike irregularity appears unexpectedly (the arrow points to it). The emission of the source at such high energy is not easily measured and therefore we do not know with certainty whether the actual energy distribution of intensity follows the curve shown here.

of the Crab nebula. Because of his earlier experience he decided to follow up on the notch.

The notch in the X-ray emission of Hercules X-1 means that at the energy level of 58 keV a particularly large number of X-ray photons is being emitted. We know that the atoms show a tendency to swallow and reemit energy at certain wavelengths, that is, at specific levels of energy. Consider the hydrogen atom. An electron revolves around the positively charged nucleus (see figure 10.13). According to quantum mechanics the electron can move only in specific, calculable tracks. When light hits the atom nothing happens as a rule, except when a photon has precisely the amount of energy required to shift the electron from an inner to an outer track. In that event the atom swallows the photon. Left to its own devices the electron in the course of time moves back to the innermost track. In doing so it emits the superfluous energy in the form of a photon that has a very specific energy value corresponding to the amount of energy released in the shift from one track to the other.

Electron in the
Electric Field

Electron in a Magnetic Field

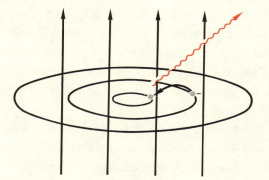

Figure 10.13 (top) Radiation in the atom (red, wavy arrow) is produced when an electron (gray) from an outer orbital around the positive atomic nucleus (red) jumps to an inner orbital. The emission has a specific level of energy characteristic for that particular atom and that transition. (bottom) In a powerful magnetic field (represented by the straight, vertical arrows) electrons can move only in narrow circular orbitals, similar to electrons revolving around atoms. In this case, too, energy is emitted when an electron from an outer orbital jumps to an inner one. The energy of the emitted radiation depends on the strength of the magnetic field. It is assumed that the notch in the distribution of X-ray intensity by the Hercules source, as shown in simplified form in figure 10.12, is caused by such jumps from orbital to orbital of electrons revolving in the magnetic field of a neutron star.

At one specific level of energy, that is, at 58 keV, the radiation emitted by the Hercules source is found to be especially strong. But no atom found in sufficient quantity in the world radiates at that level of energy. Trümper sought to explain the radiation by a mechanism first developed by the Soviet physicist Lev Landau.

The explanation is connected with the fact that in magnetic fields electrons are diverted, causing them to move in circular tracks. If the magnetic field is powerful the tracks are small. In the case of particularly powerful fields, the circular tracks are comparable to the orbits of electrons in atoms. At that point the law of quantum mechanics stating that only specific orbits are possible becomes operative. If electrons change from outer to inner tracks, the energy they emit is predetermined by the power of the magnetic field. That, according to Trümper and his co-workers, was responsible for the notch in the radiation curve of the Hercules source. If their conjecture is true the magnetic field must be 100 billion times as powerful as that of the earth. Fields that powerful would exert enormous forces, which the gravity of a white dwarf would be unable to counteract. The fields would thus shatter the star. For that reason we must assume the Hercules source to be a neutron star.

In the binary system of the Hercules source a neutron star is thus responsible for X-ray emission. At one time—and a very long time ago it was—the originally more massive star must have exploded as a supernova and left the neutron star in its wake. The explosion cloud has long since dissipated. Now mass from the originally less massive star, which is still near the main sequence, flows out to the neutron star. When that mass, steered by the magnetic field, falls down at the magnetic poles, it emits X rays. When jumping from the outer to the inner track, the electrons moving in the magnetic field in tiny circular orbits produce the additional radiation responsible for the notch at 58 keV.

Since Uhuru a number of X-ray satellites have been launched, and many balloon experiments have been conducted. One of the problems in X-ray astronomy is the fact that we have not yet succeeded in building a simple X-ray camera. X rays cannot be deflected by lenses. Mirrors also do not reflect X-ray light unless the light arrives almost parallel to their surfaces. When he developed his X-ray telescope in 1952 at Kiel, the physicist Hans Wolter (1911–1978) exploited just this property of X rays. In 1978 NASA launched the *Einstein* X-ray observatory. The main instrument was a 57-centimeter Wolter telescope that worked until 1981. Riccardo Giacconi, the father of Uhuru, and his co-workers had designed it. It found X-ray sources 1,000 times weaker than the faintest sources detected by Uhuru. A Wolter telescope of 80-centimeter diameter is supposed to be the main instrument aboard the German X-ray satellite *Rosat* scheduled for 1987.

X-Ray Showers

Still another type of X-ray source has recently been discovered. These sources, which seem to occur most frequently in globular clusters, send out showers of second-long pulses, each of which contains as much energy as the total weekly production of the sun. They do not follow the regular pattern of the Hercules source, and they seem to lack the timers of rotating celestial bodies (see figure 10.14). Still, they do come in fairly regular sequences. A globular star cluster in the constellation Scorpio occasionally sends out pulse showers in roughly 40-second intervals; the interval following a greater burst is longer than the interval after a weaker one. It is likely that in these sources matter is also sprinkled onto compact objects, but the mechanism that causes the energy thus released to be radiated in the form of showers seems to differ from the mechanism of the radiating Hercules source.

As we learned in chapter 2 globular clusters are old. No stars have been generated in them for quite some time. It was even believed that their evolution might have run its course. But the X-ray showers emanating from them are convincing evidence that there is life in them still.

Figure 10.14 The signals of MXB 1730–335 originate in a globular cluster about whose existence we first became aware through an X-ray source. The pulses come in the form of showers with ten to twenty individual outbreaks. Not all the pulses are equally powerful. After particularly powerful outbreaks the source often needs to recover before sending out new pulses. (Adapted from George W. Clark)

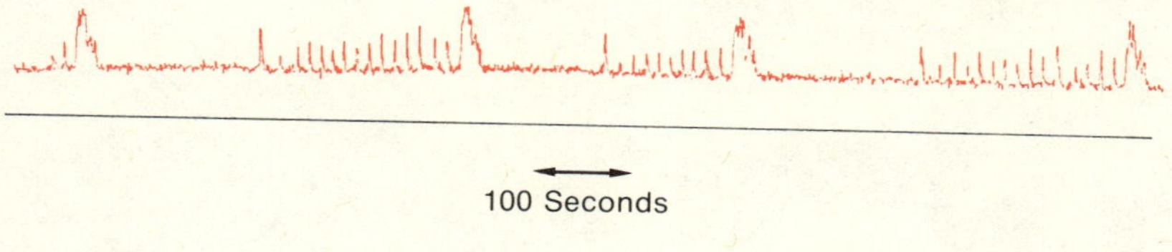

100 Seconds

It is very likely possible that there exist many more neutron stars than we are aware of. They may be remnants of supernova explosions or the products of still unexplored generative mechanisms of nature. We may never learn anything about these objects except perhaps when companion stars to which they are linked shower them with mass and stimulate them to irradiate us with X rays.

In a lecture I delivered in 1960, I asked the audience to imagine an instrument able to translate electromagnetic radiation from space into audible sound. Given such an instrument we would hear the regular hissing of the stars and the radio outbursts of the sun as well as the hissing of the known radio sources, rhythmically swelling and ebbing with the rise and fall of these radiating objects revolving in time with the vaulted heavens. All in all, I thought of it as a rather boring transmission. But now I must revise the scenario. In addition to the then known radiative sources, the ones discovered since would form part of this talking picture. Above the regular hiss we now would hear the overlapping ticking of the pulsars, the deep rumble of the Crab pulsar, whose individual pulses could not be distinguished by the ear, and interspersed with this, the rapid fire of X-ray sources like MXB 1730-355, which is sending out high-energy pulses from a globular cluster, perhaps a dozen of them in ten- to twenty-second intervals, and after a pause of some minutes a new sequence begins. Not only does the cosmos hiss but it ticks and drums, it hums and clatters. The neutron stars are probably responsible for all this noise, passed on to us from space by our imaginary receiver.

Are the pulsars and X-ray stars showing us the possible final stages in the life of a star? Do we now know that all stars end as neutron stars or white dwarfs? A strange attribute of these two types of stars leave open still another possibility.

11

The End of Stars

The velvet-black round object floated above the room. Actually the thing did not look at all like a ball but more like a gaping hole. A violent rush of air set in and continued to swell in intensity as the air in the room was sucked up into the sphere. Scraps of paper, gloves, ladies' veils—all were carried along by it. And when a soldier pushed his saber into that eerie hole, the blade vanished as though it had melted away.

Gustav Meyrink,
Die Schwarze Kugel, 1913

The pulsars and X-ray sources have brought us knowledge of neutron stars. One of them appeared in the wake of the Crab supernova explosion in 1054. But what caused that explosion? If we were able to witness another such explosion in our galaxy, we might find out what actually is exploding, for we could look at old photographs to find the star that has been shattered into a cloud in whose interior a small neutron star is spinning like a top.*

In the absence of eyewitness reports we must settle for speculation. We can once again consult the computer models of highly developed stars and ask whether they can predict how a star arrives at this catastrophic turn of events.

*Perhaps we ought to be a little cautious about wishing for a close view of a supernova. According to Malvin A. Ruderman of Columbia University in New York, a supernova explosion at a distance of fewer than 30 light-years could well be fatal for us humans. The radiation of high-energy particles released by it would destroy the ozone barrier of the earth's atmosphere; ultraviolet radiation from the sun would come down to us unfiltered and ultimately put an end to life on Earth.

The Iron Catastrophe in Massive Stars

Massive stars, that is, stars of more than 10 solar masses, evolve very rapidly. Their hydrogen supply is depleted after only some millions of years. At that point the helium ignites and is transformed into carbon, and before long the carbon atoms change into higher atomic nuclei. During all these nuclear reactions energy is released, but the nuclear processes become increasingly more unproductive. They have to continue to accelerate to cover the undiminished stellar radiation. Increasingly complex atoms are built up in rapid succession. How long does this process continue?

Nature has put a limit to it in the chemical element iron. We have already mentioned that nuclear reactions involving higher elements deliver less and less energy. With the atomic nucleus of iron the nuclear reactor in the star is extinguished. When the iron nuclei in a star fuse with other nuclei in that star, no energy is produced. As a matter of fact the process requires energy. Smashing the iron nucleus also requires energy.

This is connected with the properties of atomic nuclei. Heavy nuclei like those of uranium give off energy when smashed, and hence turned into nuclei whose weight is closer to that of iron, a lighter element. Light nuclei like hydrogen and helium give off energy when they are fused into new nuclei whose weight is closer to that of iron. But iron itself provides no nuclear energy.

What happens when the fusion of the individual elements in our massive star has progressed so far that its central region resembles a sphere of gaseous iron (see figure 11.1, left)? The atomic nuclei of iron can capture the electrons buzzing around in the gas. The iron sphere must then shrink, for its gravity and gas pressure are in equilibrium. The electrons are primarily responsible for the gas pressure. When electrons disappear in atomic nuclei, gravity gains the upper hand over gas pressure. The sphere of gaseous iron in the central region of the star then collapses into itself. This process sets in when the iron sphere is about 1.5 solar masses and it continues until, at

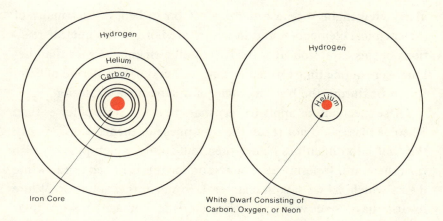

Figure 11.1 Possible preliminary stages of a supernova explosion. (left) A star of 10 or more solar masses. In its interior heavier elements have formed out of the original hydrogen-rich matter, which continues to make up its outer layer; these elements form concentric layers all the way to the core, where the matter consists of highly compressed iron gas. The central area of this evolved star is not stable; it can implode. In the process so much energy is released that the layer surrounding the central region can be hurled into space. (right) The highly evolved stage of a less massive star. In its interior is a core composed primarily of carbon with the essential characteristics of a white dwarf. The mass of the white-dwarf core increases because helium burns on its surface and changes into carbon. When the white dwarf reaches Chandrasekhar's limit of 1.4 solar masses it collapses, and the energy released hurls the outer layers into space. The two drawings are schematic; they are not true to scale.

extremely high densities, all nuclear building blocks have moved so close together that ultimately all protons and electrons are fused into neutrons. All that is left at that point is neutron matter. The dense, gaseous iron sphere in the star's interior has turned into a neutron star. In the course of this transition an unfathomable amount of energy is released that probably hurls the outer layer of the star into space at high velocity. The star explodes, leaving a neutron star in the middle of the scattering explosion cloud. The star has ended its life in a supernova explosion.

At various research centers—in Chicago, in Livermore, and in Munich—efforts are underway to simulate this process with computers. The calculations for this experiment are far more difficult than those for the normally slow evolutionary phases. But they are particularly challenging because there is reason to believe that many of the chemical elements found in nature are born in the nuclear reactions of this explosion. It is likely that all elements heavier than helium were at one time formed in stars, either during a period of calm fusion or during the brief moment of a supernova explosion.

These speculations apply to very massive stars. Stars of fewer than 10 solar masses do not reach the iron phase in their fusion process; they get into difficulties before then, difficulties which possibly also turn them into supernovae. The reason for this is linked to the white dwarf which, as we saw in chapter 7, forms in their interior. White dwarfs have one very odd characteristic connected with their equilibrium.

An Imaginary Experiment Involving a White Dwarf

We owe our existence to the balance of gravity and pressure in the sun and the earth, and generally we can rely on the maintenance of this balance. If in an imaginary experiment we were to squeeze the sun just a little, its gravity would increase because its matter would have moved closer together, but its interior pressure would also go up, even more than gravity, and the sun would return to its original state of equilibrium. Conversely, if for some reason the sun were to be so greatly deformed by some force as to temporarily increase in volume, its gravity would decrease, since the mutually attracting particles of matter would have moved farther apart. Yet at the same time pressure would decrease, even more than gravity, causing the sun to again return to its original state. We have said that we can rely on the equilibrium of the sun. Translated into scientific terminology this means that its equilibrium is stable. But not all stars are this reliable. White dwarfs are stable, but they are easily destabilized.

Long before we knew how stars evolved, years before anything was known about the stellar nuclear reaction that changes hydrogen into helium, and long before there was any inkling that stars could be simulated in computers, a twenty-four-year-old Indian scientist at Cambridge, England, solved the equation describing the structure of a white dwarf. Subrahmanyan Chandrasekhar, born in 1910 in Lahore, India, had been an outstanding student at the University of Madras. He entered an essay contest and won as a prize a copy of Eddington's book on the internal constitution of the stars. This may well have turned his interest toward stellar structure. Among his many contributions to astrophysics was his proof that white dwarfs cannot be composed of an arbitrary amount of matter. I will try to demonstrate this thesis with the help of an imaginary experiment.

Suppose human beings were tall enough to reach up to the stars. We would then be able to take away mass from one star and give it to another. Suppose also that we are floating somewhere in the neighborhood of Sirius where the white dwarf Sirius B is revolving about Sirius A. Sirius B, having less than 1 solar mass, is small compared with the sun. Its radius is only seven one-thousandths of a solar radius. Now suppose we have a great supply of white-dwarf matter and are intent on increasing the mass of the white dwarf by slowly pouring that matter onto its surface. We would find that as we increase its mass, it shrinks in size. By the time we have added enough mass to reach 1.33 solar masses, its radius will have shrunk to four one-thousandths of a solar radius. However carefully we continue to add mass, the shrinking accelerates with increasing mass. The pressure in the interior of the white dwarf is less and less able to keep in balance with gravity. The star continues to contract, and because gravity continues to increase things go from bad to worse. At 1.4 solar masses, gravity gains the upper hand and the star becomes imbalanced. This critical mass is known as *Chandrasekhar's limit.* Once that limit is exceeded the structure collapses within seconds. The density of the electron and helium gas increases and soon a familiar process sets in: the electrons that come into the vicinity of atomic nuclei penetrate them, turning the protons in them into neutrons, and the atomic nuclei disintegrate. What is left are neutrons, which now form the collapsing matter, and in the process the velocity of the collapse accelerates. The neutrons fly toward the center at a high rate of speed. When the matter has shrunk to a radius of 10 kilometers, the pressure of the neutron gas again becomes great enough to put a brake on gravity. The shrinking stops; the matter is halted in its motion. The energy of that motion is radiated outward, and what remains is a body in equilibrium. Because it is composed primarily of neutrons, it is called a *neutron star.*

This imaginary experiment, in which we have arbitrarily added matter to the white dwarf, is not all that far-fetched. We know that white dwarfs form inside red giants and that they consist of matter that has already gone through hydrogen, and possibly also helium, fusion. On the surface of these white dwarf-like cores hydrogen is still being transformed into helium. Increasingly greater amounts of the outer unexpended matter go through hydrogen fusion, and probably helium fusion as well, and are incorporated into the red giant's dense white-dwarf core. The white dwarf gains mass. As in our imaginary experiment more and more mass is channeled to it (see figure 11.1, right). What happens then, after it exceeds 1.4 solar masses, that is, after reaching Chandrasekhar's limit? It would then begin to shrink and collapses into a neutron star.

Some researchers believe that no such star ever reaches the neutron stage because a carbon explosion is likely to take place before that happens. As yet very little is known about any of this. Let us assume that the white-dwarf core of our star is composed primarily of carbon. That carbon will most likely ignite before the collapse, and in exploding tear the star apart before a neutron star has formed. In such a supernova no neutron star would inhabit the explosion cloud; no pulsar signals from it would come down to us. And in fact no pulsar has been found at the site of either Tycho de Brahe's or Kepler's supernovae, even though both these clouds are of more recent origin than the Crab nebula. The Einstein Observatory orbiting the earth has measured the remains of a supernova in Cassiopeia that erupted only three hundred years ago. Because the explosion took place behind absorptive dust clouds, it was not seen by us. The cloud apparently does not contain a neutron star. Is it possible that a star died there in a carbon explosion?

Do all our less massive stars end in carbon explosions? Nobody knows for sure. It is altogether possible that after igniting, the carbon burns off comparatively harmlessly without shattering the star. In that event the white dwarf in its interior would continue to increase in mass until it reached its limit and, as in our imaginary experiment, collapse into a neutron star. As in the iron catastrophe, the energy released in the process would certainly suffice to bring us the awesome drama of a supernova explosion. Perhaps the supernova of 1054, which gave birth to the Crab nebula, was the final stage of just such a process. Perhaps its story is as follows.

There once was a 5-solar-mass star that burned hydrogen at its center, and after depleting the nuclear fuel in its interior turned into a red giant. The helium in its core ignited and was used up, and the star acquired a carbon core. In the stellar interior there was then a helium shell surrounding the carbon core, and the matter in the core had the density of a white dwarf. Nuclear burning on the surface of the helium mass turned hydrogen into helium, and at the border between helium and carbon, helium was turned into carbon. The mass of this core, actually a white dwarf, continued to increase while at the same time its radius decreased, until in 1054, when it had reached about 1.4 solar masses, it collapsed. Even carbon burning was no longer able to halt the collapse. Enormous amounts of energy were released, and in the explosion the outer layer was scattered. That outer layer is what we today see as the Crab nebula. Within seconds of the explosion the white dwarf changed into the neutron star that is sending out the signals of the Crab pulsar.

Which of these three scenarios is the one responsible for the supernova explosion? The iron core in the stellar interior that collapses into itself in an implosion? The white dwarf that, like a cancerous growth, eats up more and more of the stellar matter until it exceeds its critical mass and also implodes? Or the carbon detonation that tears the star apart before the white dwarf can turn into a neutron star? In other galaxies we find two kinds of supernovae. They differ in their light emission. Perhaps all of the conjectured mechanisms are at work, different ones in different kinds of supernovae. Perhaps more massive stars die because of their iron core, whereas stars ranging from 10 to 1.4 solar masses die because of the white dwarf growing in their interior—be it through carbon detonation or the formation of a neutron star.

Only stars of fewer than 1.4 solar masses and others that rid themselves of superfluous mass with the help of stellar winds or the blowing off of planetary nebulae can look forward to a peaceful death. They become white dwarfs in which nuclear reactions no longer take place and equilibrium is maintained.

An Imaginary Experiment with a Neutron Star

Neutron stars also have a problem with equilibrium. Let us conduct another imaginary experiment. Let us take the Crab pulsar, which possibly consists of 1 solar mass of neutron matter, and let us assume that we can, in a space experiment, add to the mass of the neutron star by carefully pouring neutron matter onto its surface. Here again the star's radius would shrink as its mass increases—a sign that gravity is gaining supremacy over pressure. When this concentrated body grows to approximately 2 solar masses, it will suddenly collapse still more, a process that takes no more than fractions of a second. Can it be stopped? Can matter change into a new type of material whose pressure will again be strong enough to balance gravity, just as the white-dwarf matter changed into neutron matter and reestablished equilibrium? Physicists believe that nothing can save the collapsing neutron matter if its total mass exceeds about 2 solar masses.

Gravity will continue to build up, and before long pressure will no longer play any role; the star will continue to collapse. In the vicinity of the collapsing mass gravity will be very great; Einstein's General Theory of Relativity deals with this natural phenomenon.

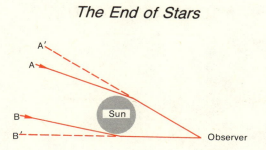

Figure 11.2 Curvature of light caused by solar gravity. Two remote fixed stars radiate light in all directions. Two light rays passing close to the sun are shown here as solid red lines, A and B. The gravity of the sun deflects the light. Seen from the earth the rays seem to come from the directions A' and B', represented by the broken red lines. To the earthbound observer the stars in the sky now appear to be farther apart than at another time of year when the sun stands in another part of the fixed-star heavens where it does not affect the light of the two stars. Because of its gravity the sun acts like a magnifying glass moving across the sky in the course of a year, enlarging the field of stars behind (insofar as they are not covered by the sun itself). The effect is minimal and measurable only during a total solar eclipse.

Among other things it describes the influence of gravity on light propagation. The sun acts like a lens on stellar light as the light passes by the sun on its way to the earth (see figure 11.2). The field of stars behind the sun appears somewhat enlarged. But the effect is very small, on the borderline of accurate measurability, and in any event it is visible only during a total eclipse when the moon covers the solar disk and the stars are visible in daylight. In the few minutes this phenomenon lasts it is possible to measure the curvature of the light rays as they pass the sun. These measurements prove that light is refracted, as posited by the theory of relativity.

The effect of gravity on the propagation of light becomes very important once the matter of our neutron star can no longer find support and collapses. Let us assume that we are able to follow this process in slow motion. At first the neutron star is still in balance, but the curvature of the light rays is already apparent on its surface since the gravitational pull is very strong. A light ray slanting upward from the surface shows a pronounced curvature until it is sufficiently far away from the area of strong gravitational pull; it then travels through space along a straight line [see figure 11.3(a)].

With the gradual increase in the mass of the neutron star, the col-

(a)

(b)

(c)

(d)

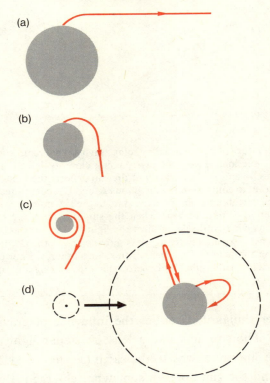

Figure 11.3 Light propagation in the neighborhood of a collapsing neutron star. (a) Near the stellar surface the light emanating from the star follows a curved course. (b) The smaller the radius of the star, the greater the curvature, so that at (c) a light ray going off at an acute angle from the surface must fight against gravity in a number of spiral windings before the star releases it. (d) The star now has become smaller than its Schwarzschild radius. Every light ray emanating from its surface now curves back toward the star. The picture to the left of (d) is drawn at about twice the scale of (c). Since the star has become very small compared with its earlier stages, (d) shows it in enlargement. The broken circles represent spheres of the Schwarzschild radii.

lapse begins because the internal pressure is no longer strong enough to balance the gravitational pull, which will become more and more powerful. Soon the curvature of the light will become so great that a ray going out horizontally from the star's surface will wind around it several times before going off into space. It becomes increasingly difficult for light to fight against gravity, and when the collapsing body, which we will assume to have grown to 3 solar masses, reaches a radius of 8.85 kilometers, nothing can escape from its surface to the outside. Every ray of light emitted is so bent by gravity that it returns to the body. The photons emitted by the body fall back down on it like pebbles tossed up in the air. No ray brings us tidings of the tragic fate of our star. Such a body is called a *black hole*.

Black Holes

We have learned that after a time a body being compressed will cease to give off light. The radial dimension at which this phenomenon sets in was first calculated by Karl Schwarzschild, director of the Göttingen and Potsdam observatories and perhaps the greatest astrophysicist of the early 1900s. He died in 1916 as the result of an illness contracted in World War I and is buried in Göttingen. Schwarzschild's pioneering contributions to astrophysics cover a wide range. Shortly before his death he found the first exact solution to the equation embodied by Einstein's Theory of Relativity, and contained in his solution is a description of the black hole.

The radius to which a body must be compressed before it ceases to send out light is called *Schwarzschild's radius.* In the case of the sun it is about 3 kilometers. If the sun were reduced to a ball of that radial size or less, no light from it would escape to the outside. Schwarzschild's radius can be calculated for every body. The smaller the mass, the smaller the radius involved. For the amount of matter of a human being, Schwarzschild's radius is so small that, expressed in centimeters, the figure would be a zero followed by a decimal point and twenty-one zeroes before any higher numeral is used. At that level of compression no light could get to the outside from the mass of a human being.

A celestial body lost in a black hole has not disappeared from the universe. Its gravity still continues to operate. It absorbs light from nearby sources and deflects passing light farther away. By dint of its gravity it can bind other bodies to itself, can keep planets around itself, and can form a binary system with another star.

All the above feats have been theoretically ascertained. Do black holes actually occur in nature? It is hard to imagine that enough matter can fall on a neutron star to destroy its equilibrium. In X-ray binaries the matter flowing to the neutron star is so paltry that during the life of the donor star the mass of the neutron star shows no perceptible increase. Still, what do we know about the genesis of neutron stars? Only that the Crab pulsar is the remnant of a supernova. And what, after all, do we know about the supernova explosion? Is it not conceivable that after the outer layer has blown off too much mass may occasionally be left over, more than the equilibrium of a neutron star can handle, and that as a result a black hole forms? There is a strong suspicion that the compact structure of some X-ray binaries sending out emission is not a neutron star but a black hole. The matter coming from the companion star may still heat up substantially before disappearing in the interior of the black hole and send forth X rays. The mass of the X-ray source can be deducted from the motion of the visible star (see appendix C), which in turn can be detected by the Doppler Effect (see appendix A). In the case of the source in Cygnus X-1, the compact structure is believed to contain more than 3 solar masses. Thus, it cannot possibly be a neutron star. Is it perhaps a black hole? But the determination of mass is not altogether reliable in this case. So we do not yet possess irrefutable evidence of the existence of black holes.

These days they keep on turning up more frequently both in scientific as well as popular literature than in nature. It has become rather fashionable to use black holes to explain phenomena that defy more conventional explanations and to hold them responsible for a vast array of puzzling cosmic occurrences. While browsing through a bookstore in London, I came across a volume on black holes on the shelf labeled *Occult Literature.* I could not help but feel that the bookseller obviously had a good feeling for this aspect of modern astrophysics.

It appears that a star is fated to end either as a trusty, cooling white dwarf or as a neutron star that in its early stages sends out radio impulses and then, if for any reason matter rains down on it, calls attention to itself by turning into an X-ray star. If, at the end of a star's evolution, a great amount of matter is left over, too much to make a white dwarf and too much for the equilibrium of a neutron star, this remnant is fated to collapse forever and ever in a black hole.

Stars end their lives as compact structures in which matter remains bound together forever, although prior to their demise they emit a portion of their mass into space. This mass becomes available for the formation of new stars. The matter of which our human bodies are made up was beyond any doubt at one time a seething mass in the interior of a star and somehow has escaped from that place and from the fate of the rest of the star's matter. In almost every case stars turn into compact bodies, and perhaps ultimately all matter in the cosmos will condense into cooling white dwarfs, neutron stars, or black holes orbited by listless cooled-off planets. The universe appears to be heading toward very boring times.

12

How Stars Are Born

We have followed the lives of stars from the ignition of hydrogen in their early days to their old age. But what happened before their birth? Where did the stars whose lives we followed come from? How did they originate, and out of what?

Since the life of a star is finite, it must have originated at a finite time in the past. How can we find out something about that formative process? Is there a place in the sky that can tell us something about it? Are we witnesses to their birth? The flat disk of our galaxy is made up of hundreds of billions of stars. Does any one point give us an indication of their origin?

Stars Are Still Being Born

Our accumulated knowledge about the life of the stars holds the key to the solution. As we have seen, massive stars—that is stars of 10 and more solar masses—age rapidly. They manage their hydrogen supply carelessly and do not stay on the main sequence for very long. So when we see a massive main-sequence star, we know that it cannot be terribly old. We recognize it by its great luminosity; because of its high surface temperature it gives off a blue light. Bright blue stars are hence young stars, scarcely more than a million years of age, youthful compared with the billions of years the sun has been shin-

206

ing. Bright blue stars mark those places in the sky where stars have recently come into being and for all we know may still be forming.

Nests of bright blue stars are found throughout the heavens. What are their striking features? Do they tell us anything about how stars originated? In the majority of the cases high concentrations of interstellar gas and dust are found at their sites, as for example in the Orion nebula (plate VI). Imbedded in it are bright blue stars one million years of age at most. In the constellation Sagittarius the young stars are obscured by a dense cloud of dust. Only through observation under longwave infrared light was Hans Elsässer and his colleagues of the German Spanish Observatory at Calar Alto able to photograph the newly formed stars through the dust and to study them.

As we know, interstellar space is not empty. It is filled with gas and dust masses. The gas density is about one hydrogen atom per cubic centimeter, and the temperature hovers in the vicinity of $-170°C$. The interstellar dust, at $-260°C$, is considerably cooler. But at the site of youthful stars, the interstellar matter is different. Dark dust clouds obscure the light of the stars behind them. Gas clouds are luminous, heated to temperatures of up to 10,000° by adjacent youthful stars. The clouds have densities of ten thousands of atoms per cubic centimeter. Complex molecules like formic acid and alcohol radiate in their characteristic radio wavelengths. The concentration of interstellar matter at these sites suggest that stars are formed out of interstellar gas.

This assumption is also supported by an idea first advanced by the English astrophysicist James Jeans, a contemporary of Eddington. Let us imagine space filled with interstellar gas in which each atom exerts gravitational pull on the rest and in which the gas is compressed. As a rule the gas pressure prevents collapse. The equilibrium maintained is very similar to that found in the interior of a star, where gravity and gas pressure keep in balance. Now let us take a specific amount of interstellar gas and in our mind's eye squeeze it together slightly. When compressed the atoms move closer and the gravitational pull increases. But so does the gas pressure, generally more than the gravity; the gas thus squeezed together again expands and is returned to its original state. The interstellar gas is said to be *stable*. Jeans discovered that this stability is not all it's reputed to be. If a large enough amount of matter is compressed simultaneously, gravity increases more rapidly than gas pressure, and the mass continues to shrink. This means that if any condensation of matter takes place at all it is sudden and involves a large mass impelled by its own gravity. About 10,000 solar masses of interstellar matter become *unstable* simultaneously. Perhaps that is why young stars are found only in groups. They are without exception born simultaneously in big litters. While 10,000 solar masses of interstellar gas and dust are collapsing into themselves at high rates of speed, cloud fragments are probably forming in the gas, each of which again condenses and later becomes a star.

A Star Is Born in the Computer

A young Canadian astrophysicist, Richard B. Larson, described the birth of a star in his doctoral dissertation, presented in 1969 at the California Institute of Technology. Larson's dissertation is a standard work of modern astrophysics. In it he explored the origin of a single star in interstellar matter. His calculations offer detailed observations on the fate of an individual cloud.

Larson assumed a spherical cloud of exactly 1 solar mass and followed its evolution in a computer program simulating the collapse of a gas cloud as precisely as was then possible. He did not start with matter in the state of the interstellar medium but with an already condensed smaller cloud, that is, the fragment of a collapsed bigger mass. Consequently, the cloud is denser than the interstellar medium; it contains 60,000 hydrogen atoms per cubic centimeter. Larson's cloud measured about 5 million times the diameter of the sun that would eventually form from it. From there on everything happened very quickly, astrophysically speaking, that is, within the span of 500,000 years.

Initially, the gas is transparent: each dust particle radiates light and heat. This radiation is not held back by the surrounding gas; rather, it goes out into space unimpeded. This early transparent model defines the future of the gaseous sphere. Gas flows toward the center in a free fall, and as a consequence matter accumulates in the central portion. The initially homogeneous mass turns into one that grows denser toward the center (see figure 12.1). That makes for accelerated gravitational pull at the center, and the velocity of matter also continues to accelerate, above all in the interior. In the beginning almost all hydrogen is joined into hydrogen molecules: two hydrogen atoms are bound to each other and together form a molecule. At first the gas temperature hardly goes up. The gas is still so rarified that all radiation can escape to the outside, and the contracting gas mass does not heat up perceptibly. After a few hundred thousand years the central region has so increased in density that its gas has become opaque to the radiation that had hitherto dissipated the heat. As a result, a core heats up in the interior of the big mass. It measures one two-hundred-fiftieth in diameter of the shrinking cloud that contains all the falling matter. The pressure rises along with the temperature and ultimately puts a stop to the contraction. The radius of the condensed core is almost that of Jupiter's orbit but contains only 0.5 percent of the mass involved in the collapse. Matter continues to rain down onto the small core in the interior, carrying energy that it emanates at impact. At the same time the core contracts and heats up.

Everything goes along well until the temperature reaches about 2,000°. At that point the hydrogen molecules begin to break up and turn back into atoms. This process has a crucial effect on the core, which once more begins to collapse and continues to do so until the energy released during the collapse has converted all the molecules into atoms. The newly formed core is only a little bigger than our sun. Finally, all the matter on the outside will fall down onto this central core and a 1-solar-mass star will result. In effect, from now on only this core is of any significance.

Because the core gradually will turn into a star, it is called a *protostar*. Astrophysicists use this term for the cores of shrinking clouds of interstellar matter whose chemical composition is identical to that

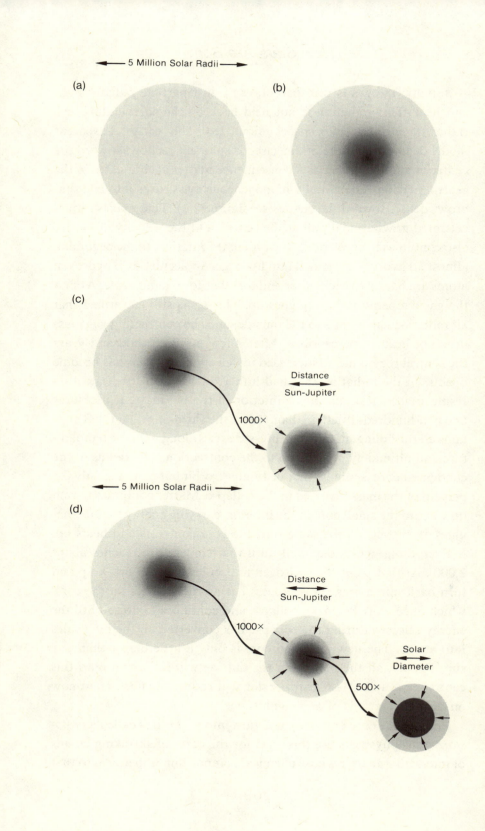

← 5 Million Solar Radii →

(a)

(b)

(c)

1000×

Distance
Sun-Jupiter

← 5 Million Solar Radii →

(d)

1000×

Distance
Sun-Jupiter

500×

Solar
Diameter

of the cloud out of which they had formed. Later, after all the matter from the cloud has fallen onto their surfaces, they will ignite their hydrogen in the central region. In the H-R diagram we will then find them on the initial main sequence. They are *initial stars*. But let us go back to the protostar at a much earlier stage, at its stage of formation in the center of the shrinking cloud. It covers its radiative needs primarily through the matter raining down onto it. Density and temperature increase, and the atoms lose their electron layers; they are said to *ionize*. Comparatively little can be seen from the outside. The protostar is surrounded by the dense layer of the gas and dust masses falling down onto it, which act as a barrier to all visible radiation. The protostar illuminates the gas and dust layer from the inside. Only after more and more of the falling matter has fused with the core does the layer become transparent, and the star breaks out into visible light. As the rest of the cloud rains down on the star, it condenses, which raises its internal temperature until the center reaches 10 million degrees and the hydrogen ignites. At that point the collapsing 1-solar-mass cloud has become a normal main-sequence star: it has turned into the initial sun, and from here on the story follows the script outlined in the beginning of the book. In the last stages of the protostar period, before the star reaches the main sequence, energy is transported through convection over wide areas of its interior. Before it gets to the main sequence in the shape of the initial sun, the solar matter is once more forcefully agitated. This finally solves the lithium problem of the sun mentioned in chapter 5. The

---

Figure 12.1 Larson's model of the origin of the sun. (a) A cloud of interstellar gas begins to collapse. At first its interior density is uniform throughout. (b) After 390,000 years the density in its center has increased a hundredfold. (c) 423,000 years after the onset of the process, a hot core (here shown magnified) that at first does not collapse onto itself develops inside the collapsing cloud. Its density is 10 million times that of the initial phase, but most of the mass is still found in the surrounding collapsing cloud. (d) Shortly afterward, when the hydrogen molecules in the core dissociate into individual atoms, the core collapses once again into itself and forms a new core, which (magnified twofold in the illustration) has attained the approximate size of the sun. For the time being it does not possess much mass, but eventually the matter of the cloud will fall on it. At that point the center of the core will become hot enough for hydrogen to ignite, resulting in the birth of a 1-solar-mass main-sequence star.

atoms of this fragile element are carried to the inside by convection, into hotter regions, where in the process shown in figure 5.3 they are converted into helium atoms before the star's change into a main-sequence star. The lithium that the solar matter received from the presolar cloud is destroyed during the protostar phase.

A Star Is Born in Nature

This is how Larson's calculation, carried out with all the imagination essential for the solution of a computer problem, ends. But does the process it describes conform to reality? Does nature follow the same pattern as the computer? Let us go back to the sites in the sky where stars are still being born, back to the young, bright blue stars, and there search for traces of the origin of stars, for objects we may expect to find on the basis of Larson's calculations.

Bright blue stars are extremely hot; their surface temperature lies in the 35,000° range. Accordingly, the radiation emitted by them is very high in energy. Their photons can snatch the electrons from the hydrogen atoms of the interstellar medium, leaving only the positive atomic nucleus. The hydrogen becomes ionized. Bright massive stars ionize the surrounding gas masses. These areas of our galaxy catch our eye by the luminosity produced when the ionized hydrogen atoms retrieve their electrons and in the process emanate light. Their luminosity makes the areas of ionized hydrogen conspicuous. Their radiative heat can also be measured in the radio range.

The radio measurements have the advantage of not being tainted by absorbing dust masses. The loveliest example of a place in the sky in which bright massive stars make interstellar matter luminous is our old friend the Orion nebula (see Plate VI). Does it contain formations that are in any way related to Larson's calculated processes? What should we look for? For extended periods the protostar is hidden by the dust layer that is slowly raining down onto it; the dust absorbs the radiation emitted by the core, that is, it receives energy. In the course of this absorption the temperature of the dust layer rises hundreds of degrees, and its radiation intensifies accordingly. Logic would dictate looking for this heat radiation in the infrared range.

In 1967 Eric Becklin and Gerry Neugebauer of the California Institute of Technology at Pasadena discovered an infrared star in the Orion nebula with a radiative temperature of 700°, a luminosity equal to 1,000 solar luminosities, and a diameter of about 1,000 solar diameters. That is exactly what the gas-dust layer of a protostar would look like. In recent years evidence has begun to pile up that compact radiative sources exist in the infrared as well as in the radio range in regions of our Milky Way in which new stars may still be expected to form. Accordingly, radio astronomers have discovered sites of great hydrogen concentrations in the Orion nebula that send out particularly strong radio emissions. In them the number of free electrons per cubic centimeter not bound to hydrogen atoms is about 100 times as great as in the Orion nebula itself. Compared with the dimensions of the entire Orion area, these bodies are very small, an estimated 500,000 solar diameters, about one-fourth of the cloud that rained down on the core of Larson's model.

The Orion region also contains bodies of lesser diameter containing radiating molecules, particularly those of water. They radiate in the radio range and can be detected by radio telescopes. These, too, are located in small, compact areas of only about 1,000 solar diameters. As we know, Larson's cloud originally had a diameter measuring millions of solar radii. The radio emission of the molecules must come from the core area.

However, caution should be exercised in venturing interpretations. The only thing that is certain is that objects have been observed in the region of the Orion nebula possessing the kind of strong concentrations of gas and dust one would expect to find in Larson's cloud, even though they are not visually striking. Yet other factors support the assumption that the concentrated radio and infrared sources really are protostars. At our institute a team working with the Austrian astronomer Werner Tscharnuter recently repeated Larson's calcula-

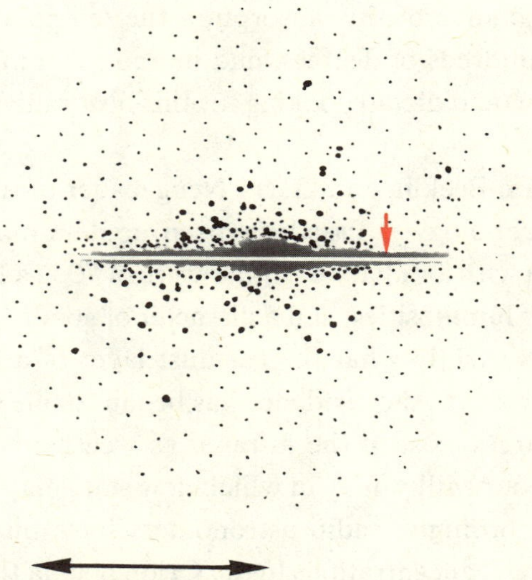

100,000 Light-Years

Figure 12.2 The schematic structure of our Milky Way system. Most of its stars are in a flat disk, here seen edgewise. The arrow indicates the location of the sun; the light central strip represents the absorbing dust masses. The globular clusters (larger dots) and very old stars (small dots) form the halo of the Milky Way. The halo stars were born a long time ago. Today stars continue to form only in the immediate vicinity of the central plane amidst the dust masses.

tion, using improved techniques. These scientists also calculated the radiative processes in the infrared range, and it would appear as though in nature we were really seeing the protostars simulated by the computer.

Since we seem to be so close to the solution to the question of the origin of stars, we must ask ourselves whether the explanation applies to all the 100 billion stars of our galaxy. Figure 12.2 offers a schematic view of the structure of our stellar system. Not all the stars are found in the disk; the oldest ones are located in an almost spherical space called the *halo*. These halo stars are old, as can be seen in the H-R diagrams of the globular clusters in that area. Their atmospheres show the composition of the interstellar matter of the time they were formed. Compared with the sun they therefore contain a smaller amount of heavy elements (heavier than helium), sometimes only one-tenth as much. The stars in the disk contain a greater amount of heavier elements. Although the amount of elements heavier than helium is still comparatively small, these stars do give us important clues for the decoding of the story of our Milky Way. Hydrogen and helium have existed since the beginning of time. They are, so to speak, God-given. The heaver elements, on the other hand, must have originated at a later time, in stars and in supernova explosions. The chemical differences between halo and disk stars are thus related to nuclear reactions in stars.

At the beginning of our galaxy or shortly thereafter, halo stars formed out of matter that had almost no heavier elements. The *massive stars* evolved first and produced supernovae that polluted the interstellar gas with heavy elements. The stars of this generation, which had *low mass,* evolved very slowly. Their outer layers (and the central regions of most of them) even today contain no heavy elements. After the halo stars were formed and after the more massive among them had already exploded, new stars were formed out of the interstellar gas, by now enriched with higher elements. These stars were born after the halo stars and are thus younger. On the other hand their outer layers contain a larger share of heavier elements than the atmospheres of halo stars. This explains why the old stars in our galaxy are poorer in heavy elements than young stars. The most recently formed stars have the highest content of heavier elements in their outer layers since they were formed out of interstellar matter that has been polluted throughout the history of the galaxy.

But why do we find the old stars that are poor in heavy elements in the halo, and the stars rich in heavy elements in the disk? Today we believe that we understand the laws governing the structure of the Milky Way. In this connection let us briefly go back to some of the things we were taught in our elementary physics classes.

Angular Momentum and Collapsing Clouds

The description of the physical world is substantially simplified by a number of conservation laws. We rely on them in our everyday lives without even being aware of them. We probably remember learning about the conservation of mass and of energy; both laws find application in our daily lives. We may not be quite as familiar with the fact that the angular momentum of a rotating body left to its own devices cannot simply disappear. Yet all of us are familiar with an illustrative example. When a figure skater does her pirouettes, she begins by rotating slowly with arms extended. As she pulls her elbows closer to her body, her rotations accelerate—the result of the conservation of angular momentum. This becomes clearer if we look at a rotating cloud, though it is not quite as graceful as the skater. The cloud may rotate about itself perhaps once in 10 million years. When it subsequently collapses into itself—let us say to one-tenth of its original diameter—it will rotate 100 times as rapidly, that is, it completes a rotation every 100,000 years. If it gets smaller still, it will rotate even more rapidly. Roughly, the law of the conservation of angular momentum might be put thus: during the collapse the number of rotations per unit of time multiplied by the surface of the more or less spherical cloud will always be the same. If the cloud becomes smaller it rotates more rapidly.

In the process, the centrifugal force increases, and at the equator of the rotating cloud it counteracts gravity: the collapsing cloud flattens out. This has consequences for the origin of individual stars, but it also bears on the origin of the Milky Way.

On the Track of the History of Our Milky Way

The origin of the Milky Way is a mystery. Some time or other a cloud of about 100 billion solar masses must have detached itself from dispersing matter originating at the beginning of time and increased its density. Like all matter this gas evolving out of the turbulent moving matter was endowed with angular momentum. It gradually collapsed, became sufficiently dense to allow pockets of clouds to form, which subsequently split up into smaller, more concentrated gas clouds. The first stars were thus born. They still consisted solely of hydrogen and helium and consumed their hydrogen according to the proton-proton chain. But soon the most massive of them used up their nuclear fuel, exploded into supernovae, and infused the gas masses with elements heavier than helium. All this happened when the entire galactic cloud was still almost spherical in shape [see figure 12.3(a)]. That is why the oldest stars and very old star clusters are found in the spherical halo of our galaxy. The stars in the halo were the first ones; they were born long before the Milky Way was a disk, long before there was a sun. In them the heavier elements are still very sparsely represented.

This sounds surprising since one would expect the old stars to have had sufficient time to produce heavy elements. This is partially true. The rapidly evolving massive stars among them produced heavy elements and dispersed them in the galaxy via supernova explosions. But these stars are gone; some may have become pulsars. The old halo stars that we see today are of low mass and are evolving very slowly. Some of them have not yet finished the hydrogen burning begun during the last 10 to 15 billion years. They are still on the main sequence and far from producing higher elements. Their fate may be that of our sun. In future they may undergo violent helium burning (helium flash) and become white dwarfs, for the rest of their life shamefully hiding the few elements heavier than helium (like carbon and oxygen) they produced in their very central regions. They will never reach the temperatures required for higher nuclear reactions. The chemical composition of the outer layers of these stars is poor in heavy elements. They have never polluted the interstellar gas with higher elements nor will they do so in future.

But the evolutionary process of the galaxy continued. The interstellar gas acquired more and more mass containing higher elements formed in massive stars and brought into the interstellar space by

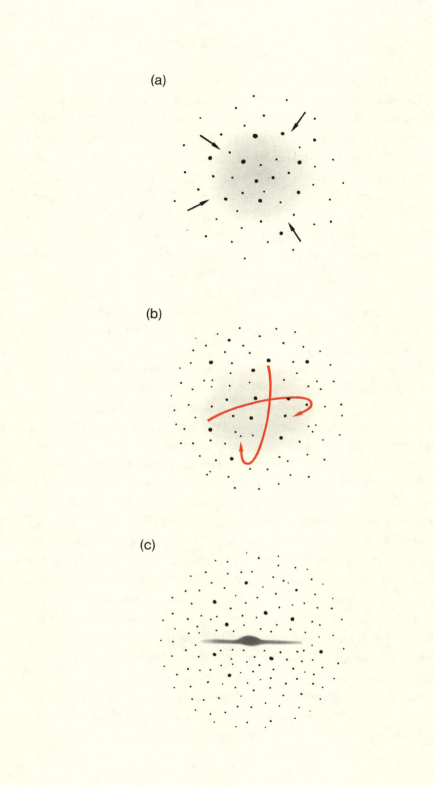

supernova explosions. The atoms of these higher elements settled on condensing solid particles blown off by evolved stars, and formed dust grains. Before long, the rotary motion of the galactic gas became noticeable. The concentrating gas-dust masses flattened, leaving the already formed stars and star clusters behind in the spherical halo [see figure 12.3(b)]. New stars of an increasing concentration of heavier elements formed only in a flat and flattening lens of matter. Most of the gas was used up when the last stars finally were born in a disk. The first phase of star formation had come to an end.

This panoramic view of the origin of the Milky Way was developed in 1962 by Olin J. Eggen, Donald Lynden-Bell, and Allan R. Sandage in Pasadena. In the twenty years since it has not lost its appeal, for it explains the essential properties of our galaxy: the oldest stars are found in the spherical-symmetrical halo deficient in heavier elements. The youngest stars are found in a thin disk, for only there is gas still present.

The angular momentum of the cloud accounts for our disk-shaped stellar system. It also accounts for the ribbon of the Milky Way we see in the sky.

What Triggers the Birth of Stars?

What causes interstellar matter to condense in specific areas of the Milky Way and form stars? Why do stars not form in other parts of the system? Seen from outer space the Milky Way would look like

---

Figure 12.3 Schematic representation of the birth of our Milky Way system. About 10 billion years ago a cloud fragment of the original matter of the universe collapsed onto itself through the force of its own gravity. With increasing density (a) the first stars (dots) and the globular star clusters (heavy dots) formed. They continue to fill the globular space in which they originated and follow the course around the center shown in red in (b). The more massive stars go through their evolution rapidly and return matter enriched with heavier elements to the gas; now new stars enriched with heavier elements can form. With the increasing density of the gas, rotation becomes noticeable; the gas forms a disk in which stars continue to form to this day (c). The schematic drawing explains the spatial structure of our system and the chemical difference between halo objects and stars in the disk.

the Andromeda nebula: a flat disk with a pronounced spiral structure (see Plate I). Other stellar systems have an even more pronounced spiral structure (see figure I.1). The striking appearance of the spiral arms in the pictures of remote stellar systems results from the presence of ionized hydrogen, which intensifies their visual brightness. As we know from the Orion nebula, hydrogen is ionized by bright, massive main-sequence stars. Thus, the spiral arms are the sites at which young stars congregate, sites at which stars have just come into being. In the Milky Way, too, we find young stars aligned along the spiral arms.

The techniques of radio astronomy have enabled us to obtain precise data on the distribution of interstellar gas in the Milky Way, and we subsequently found out that the gas is more densely concentrated along the spiral arms than in the rest of the disk. Not only are spiral arms sites of greater gas density but they are also the areas where young stars live. The question that remains to be answered is, Where does the spiral structure that makes galaxies look like rotating pinwheels come from?

For a long time the spiral arms were a puzzle. Even today we do not fully understand them. Stellar systems rotate. Their rotational velocity can be measured (see appendix A). We also know that they do not rotate like fixed bodies. Their rotational velocity diminishes toward the outside; the galactic interiors complete their course more rapidly.

Perhaps we should not be surprised to find spiral structures in the galaxies. When we stir milk into a cup of coffee, for example, we see spiral formations because the liquid rotates faster near the center of the cup than near the rim. It is only logical to assume that because of the nonuniform rotation every galactic structure will eventually become spirallike.

Carl Friedrich von Weizsäcker once remarked that the Milky Way today would have to have spirals even if originally it had looked like a cow. Some years ago a group of us at Göttingen with the help of Alfred Behr decided to pursue von Weizsäcker's idea of a galactic cow. The result is shown in figure 12.4. Before most of the stars had

0 Years   5 Million Years

100 Million Years   200 Million Years

|—————————|
30 000 Light-Years

Figure 12.4 The Milky Way does not follow a rigid rotational pattern. That is why after the passage of 100 million years every initial structure gives birth to a spiral-form object. Unfortunately, this explanation does not apply to the spirals of the galaxy itself.

---

completed even a single rotation about the center, the cow galaxy had turned into a lovely spiral. Unfortunately, there is a catch to this.

It takes less than a 100 million years for the random initial structure to evolve into a spiral. The Milky Way, however, is 100 times as old. This means that the original structure must have formed many more spirals. Like the grooves of a long-playing record, the spirals should have been coiled around the center 100 times and more. Yet we have found no such thing. The spiral arms of a galaxy, as seen in plate II are not tightly wound, and hence they cannot be the remnant of an embryonic structure. Since none of the observed spiral systems shows very close spirals, we must assume that the spirals do not wind around. Yet they are composed of stars and gas

221

that participated in the winding rotational motion. How do we escape from this dilemma?

There is only one way out. We must discard the notion that the spirals are always composed of the same matter and instead assume that stars and gas flow through them. Even though stars and gas participate in the rotational movement, the spirals themselves merely represent a special transitional phase that stars and gas pass through.

We are familiar with a similar phenomenon from everyday life. A gas flame also is not always composed of the same matter. It is merely a particular phase of a stream of gas flowing through it between whose molecules a special chemical reaction is taking place at that site. Similarly, the spiral arms are sites in the rotating galactic disk at a particular phase of the stream of stars and gas passing through. This condition is maintained by the properties of the gravity of the matter that forms the entire galaxy. Let me elaborate on this.

What Are the Spiral Arms?

Currents in nature often produce regular forms. The interplay of wind and water creates wave patterns on the surface of a lake. And when we mix liquids of different density and temperature, we frequently find structures emerging. When a cup of hot chocolate is left to cool, regular patterns will appear on the surface. Stars in a disk moving around their common center steered by the interplay of gravity and centrifugal force also show a tendency to form structural patterns.

If we imagine a great number of stars distributed throughout a rotating disk, we find that centrifugal force and gravity maintain a balance at all parts of the disk. But in general this equilibrium is not stable. If for some reason the stars in one place are more dense, their mutual attraction is stronger, somewhat like the instability of interstellar gas that led to the formation of the stars. But now centrifugal force has become important, and consequently, the process has be-

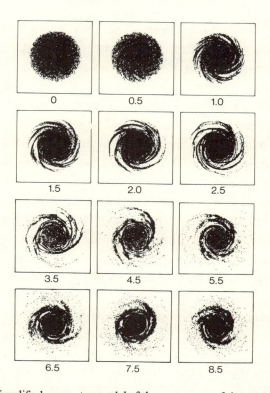

Figure 12.5 A simplified computer model of the movement of the stars in our galaxy. About 200,000 stars move around the center of a flat disk, here viewed from above. The figures below the panels indicate the time elapsed since the initial distribution in the first panel. The median orbital period was used as the unit of time. Thus, the time elapsed between the first and third panels equals one orbital cycle of the Milky Way. Stars flow through the spirals, and thus these are at all moments composed of other stars, as shown in the upper spiral arm of the panels of the 4.5 and 5.5 time frames. There one can still see how the arm of the second latter panel grew out of the earlier one. Between the two the arm rotates only minimally. But in the interim the stars had completed an orbit around the center. The model shown here was calculated by Frank Hohl, an astronomer affiliated with the NASA Langley Center, Hampton, Virginia.

come more complicated. The solution can be simulated in a computer. Figure 12.5 shows the motion of 200,000 stars in a rotating disk as calculated by computer. Long spiral filaments of greater stellar density take shape. The stars form spiral arms. The arms, however, do not coil, for they are not always composed of one and the same stars. Stars *flow through* them. When in the course of their circular orbits stars wander into a spiral arm they move closer together. When they leave again the distance between them increases. Hence, the spiral arms are places at which stars stand closer together, just as the flame is a place at which the gas molecules react chemically.

Stellar density is somewhat greater along the spiral arms than elsewhere in the disk. Even though this can be seen clearly in figure 12.5, in an actual galaxy these concentrations are so insignificant as to be imperceptible. But interstellar matter that shares in the rotational movement of the stars also increases in density when it passes through a spiral arm. This increase in density creates the preconditions for the birth of stars. That is why stars originate in the spiral arms. Among these newly formed stars are massive, bright blue ones that make the surrounding gas luminous. The luminous hydrogen clouds, not the stars closer to each other, are what make the spiral arms such spectacular sights.

We have already made the acquaintance of the galaxy in the Hunting Dogs, seen in plate II. That galaxy gives us still more information about the birth of stars in spiral arms. We see the system far out in space, shining through the nearby stars of our own galaxy. It takes light 12 million years to travel down from it to us. Since we in a manner of speaking are looking down on this galaxy from above—that is, we look down vertically on its disk—we get a particularly good view of the spiral arms.

The Birth of Stars in the Hunting Dogs Galaxy

This galaxy sends out radio signals. Rapidly moving electrons, probably accelerated by earlier supernova explosions, flow through the system and send out radio waves that can be picked up by sensitive radio telescopes. It is even possible to distinguish between those regions in the galaxy sending out more radio waves and those sending out fewer. In 1971 three radio astronomers working in Holland, Donald Mathewson, Piet van der Kruit, and Wim Brouw, made a "radio picture" of this galaxy (see figure 12.6), in which brightness

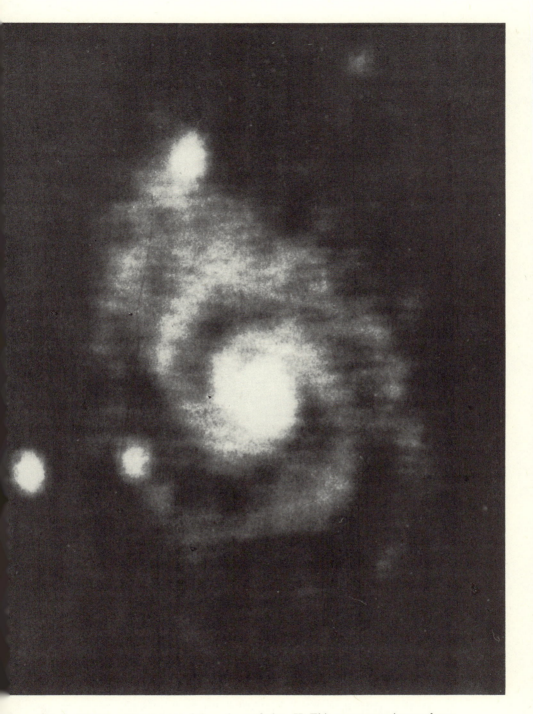

Figure 12.6 A radio picture of the galaxy of plate II. This computer picture shows the galaxy as it would appear to us if our eyes were sensitive to radio emissions of 21-centimeter wavelengths and if we could "see" as well as the big radio telescope of Westerbork, Holland. The radio emission originates primarily in regions of increased interstellar gas densities. The radio picture tells us that the spiral structure of the gas in this galaxy is almost identical with the distribution of young stars. (Photograph: Sterrewacht Leiden)

signifies radio emission: the brighter a spot, the more powerful the radio emission there. Even though the radio telescope did not see as clearly as optical telescopes, the spiral structure is easily discernible. Not only are the spiral arms bright in visible light but they also send out radio waves.

Why do the electrons in one part of a galaxy send out more radio waves than those in another part? The reason is connected with the mechanism that gave birth to this radiation. We will not go into this aspect, however. Suffice it to say that in some areas where interstellar gas is more dense, radio emission is more active. The radio picture of the galaxy in the Hunting Dogs constellation proves not only that stars in the spiral arms are closer to each other but also that the interstellar gas has greater density there.

The nebula in the Hunting Dogs can tell us still more. A closer look shows that the sites of strongest radio emission do not coincide exactly with the visible spiral arms (see figure 12.7). The greatest density of interstellar gas is found toward the inner surface of the bent spiral arms. What does that signify? The rotating galaxy flows through the spiral arms. Stars, and with them interstellar matter, move through a spiral arm in such a way that they enter the bent arm from its inner surface and later leave it by its outer surface. A comparison between the visible spiral arm produced by the newly formed stars and the radio spiral arm, which indicates the sites of interstellar gas compression, yields the following picture.

Stars and interstellar matter rotating jointly in the galaxy (see figure 12.8) approach a spiral arm. The stars move closer together, and the gas becomes denser, which creates the conditions for star formation. The first protostars are born. After a time stars and interstellar matter again leave the zone of highest density that forms the spiral arm seen in the radio range. Soon everything returns to its previous state, but not completely; the clouds that in the meantime have begun to collapse continue to do so; the forming of stars set off by temporary compression continues. After a time the first massive stars form out of the protostars. Their bright blue light makes the neighboring gas masses luminous. The newly born stars produce the spiral arm seen in visible light.

Figure 12.7 By superimposing the optical picture of the Hunting Dogs galaxy on the sites of maximal radio emission (here schematically represented by red lines), we find that the spiral arms of maximal gas density and the spiral arms of the young stars do not coincide precisely. Thus, we must draw a distinction between radio and density arms and visible arms. (Photograph: Sterrewacht Leiden)

Thus, to begin with, matter first passes through the "density arm," where it sets off the formation of stars. After a time the first stars emerge and the "visible arm" becomes luminous. Since we know the velocity of stars and gas in the galaxy, and since we can measure the distance between the "density arm" and the "visible arm," we can figure out how long after the condensation of the interstellar gas the first stars will appear—approximately 6 million years. During the last 500,000 years of that time a process similar to that described by Larson's calculation takes place in each individual cloud. It takes 5.5 million years for the cloud with which Larson began his calculation to form out of interstellar matter.

Before even a fraction of a single rotation around the center of

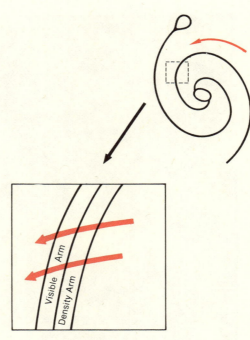

Figure 12.8 The birth of stars in the Hunting Dogs galaxy. On the upper right a schematic representation of the galaxy of plate II. The broken-line square at the upper right is shown in enlargement at the lower left. The matter of the galaxy rotating counterclockwise in the picture first flows through the density arm (radio arm), compressing the interstellar gas. Stars begin to form. After a time the first young stars appear and illuminate the neighboring gas masses, which then shine in visible light (visible arm). Since the gas moved between the onset of condensation and the birth of the star, radio arm and visible arm do not coincide. That also explains the observed difference between the radio arms and the visible arms of this galaxy, shown in figure 12.7. The movement of matter is indicated by red arrows.

---

the galaxy has been completed, the life of the massive stars has already come to an end. They have returned a major portion of their matter to the interstellar gas and have either become white dwarfs or exploded as supernovae. The matter brought to the interstellar gas, which already contains atoms of the heavier elements that originated in the stars, again becomes available for the formation of stars during the next passage through a spiral arm. Matter left over at the end of the life of a star as a compact body, either a white dwarf or neutron star, is not part of this cycle.

How Stars Are Born

At some time long after the birth of the halo stars, the matter of our sun in the form of interstellar gas also passed through a spiral arm, and many stars formed as a result. The more massive siblings of the sun have long since been extinguished, and those that like the sun were composed of less mass have since lost sight of each other, driven apart by the nonuniform rotation of our galaxy.

13

Planets and Their Inhabitants

Whether the moon is inhabited the astronomer knows as
surely as he knows the identity of his father, but not as
surely as the identity of his mother.
                    Georg Christoph Lichtenberg, 1742–1799

As a result of angular momentum, the birth of stars does not quite
follow the pattern described in the preceding chapter. Stars and interstellar matter orbit around the center of our Milky Way. Moreover, each cloud remnant formed from the gas also rotates about its
own center; this rotary motion is preserved and even intensified when
interstellar gas and dust clouds contract to form stars. This has significant consequences. As the density grows the angular velocity and
thus the centrifugal force increase, thereby counteracting gravitational forces at the equator of the cloud. The collapsing cloud flattens, ending not as the spherical protostar of Larson's calculation
but as a rotating disk (see figure 13.1). Things seem to go differently
from what we had expected.

Our solar system shows that the rotation of the original matter
played an important part in the creation of the sun. The planets all
revolve around the sun in the same angular direction; almost all their
orbits lie within a single plane, as if they had in fact been born in
a flat disk and were still exhibiting its rotation. Furthermore,

Figure 13.1 Schematic representation of the origin of our planetary system. A portion of a cloud of interstellar gas detached itself and collapsed onto itself. In doing so it flattened out, since the centrifugal force prevented inward motion in the equatorial plane. The result was a flat disk in whose center the sun formed. In the flat disk around the sun, matter condensed into planets, which then rotated in the plane of the former disk. The drawing is not true to scale. As simple as the process appears, we have not yet managed to understand it in every detail.

---

whereas most of the mass in our solar system is concentrated in its central body—the planets comprise only 1.3 percent—the sun has almost no angular momentum: most of the solar system's angular momentum resides in the orbital motion of the planets. It looks as if the matter knew just what it was doing during the collapse of the interstellar cloud when it redistributed the angular momentum that otherwise would have prevented the formation of a star. A small portion of the original mass grabbed almost all the angular momentum for itself and formed the planets, thus allowing the larger portion, now almost free of angular momentum, to form a central body à la Larson.

The Problem of Planetary Birth by Computer

The French mathematician Laplace and the German philosopher Immanuel Kant had already suspected that the sun and the planets were born of a rotating nebula. Today using computer we are in the position of simulating these processes. The following discussion is based on research carried out by the Californian astrophysicist Peter Bodenheimer and by Werner Tscharnuter, in part separately and in part jointly in Munich. Their objective was to find the key to the origin of the sun and the planets, but things did not quite turn out as planned.

The ease with which spherical-symmetrical processes can be simulated by the computer becomes apparent only as one proceeds to the next more difficult problem. In spherically symmetrical systems only the distance from the center ever matters. For one clump of matter at a given radius to heat up in Larson's model, all the matter at the same distance from the center—that is, all the matter located on the same spherical shell—must heat up simultaneously. When the matter does not rotate spherical symmetry is a good approximation to real nature, since every particle of mass involved in the collapse suffers the same fate regardless of the direction from which it comes.

Rotation, on the other hand, disturbs this symmetry. Particles flying along the polar direction feel forces different from those felt by particles coming from an equatorial direction. So much for spherical symmetry! On the other hand, at least a certain degree of symmetry remains: for example, in the equatorial plane, particles approaching the center from different directions all behave identically. Such a process is called *axisymmetric*. Although they are by no means the worst, axisymmetric processes pose a problem for the computer, though not an insoluble one. Bodenheimer and Tscharnuter followed a collapsing rotating cloud numerically (see figure 13.2). Initially, everything proceeds in line with Larson's calculation: the cloud collapses and forms a dense core. But during its collapse centrifugal force increasingly comes into play. The cloud flattens until it finally

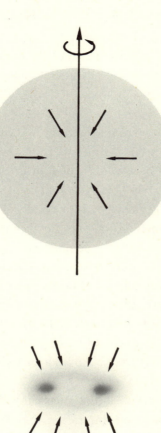

Figure 13.2 A rotating interstellar cloud collapses onto itself. The rotational axis is indicated in the top drawing. Initially, the gas falls toward the center evenly from all directions, as shown by the arrows. Later (center), a rotating disk forms, and gas flows toward it mainly from the direction of the poles. The first traces of a ring appear in the disk, shown here at the bottom in cross-section of the plane in such a manner that the ring shows up only as two condensations. A view of the same ring from above appears in figure 13.3 (top). In this process, calculated by Bodenheimer and Tscharnuter in 1978, no central star forms.

becomes a flat disk. Only the matter near the rotational axis continues to collapse, whereas the gas in the equatorial plane slowly approaches the center until it finally comes to a complete stop. Instead
of a core on which matter impinges from all sides, there is now a
disk onto which matter falls only from the axial directions. The disk,
with an equatorial radius 8 times its thickness, extends far out into
space, to about 120 orbital radii of Pluto, the most remote planet.
The disk takes about 300,000 years to rotate about its center.

That was not quite the hoped-for result. What Bodenheimer and
Tscharnuter had hoped to find was a body in whose interior the initial sun had formed. They would also have been happy with a disk
around the sun out of which planets could have formed in the course
of time. However, no solarlike body appeared in the center of the
Bodenheimer-Tscharnuter disk. On the contrary the peak density
was found in a ring at a distance of 17 Pluto orbital radii from the
center. Instead of a central body a ring had evolved! The picture in
the lower portion of figure 13.2 gives a side view of this ring. Figure
13.3 is a view seen from above.

In retrospect this result is not surprising. Why does the model's
matter form a surrounding ring rather than flow to the center? Centrifugal force prevents it from doing so. The angular momentum of
matter is the culprit. We saw earlier that in the evolution of the solar
system, nature separated matter and angular momentum, so that the
mass now belongs to the sun and the angular momentum to the planets. In their calculations Bodenheimer and Tscharnuter left every
gram of matter with the angular momentum it possessed at the beginning. It would be possible to repeat their calculations under the assumption that the angular momentum flows through matter just as
heat flows through a body. But now the following problem emerges:
we know of several possible means for transporting the angular momentum from one place in the disk to another, but we do not know
which of these mechanisms is the crucial one. Magnetic fields could
free portions of the disk from their angular momentum and thereby
allow matter to form a dense core, but turbulent movement with
frictional processes is also a possibility.

Figure 13.3 A view from above of the ring of matter shown in figure 13.2, about 3 times as large as the bottom picture shown there. After some 100,000 years the ring disintegrates into two concentrations, which may turn into two stars that later will revolve around each other as two remote binaries.

The turbulent motion of liquids and gases remains among the least understood natural processes, although its qualitative features are quite well known. Water spurting forth under great pressure from an opened faucet does not flow evenly but constantly changes its complex motion in unpredictable ways. The current of a mountain brook is another example of the turbulent, irregular motion of liquids. That turbulence can play an important part in the rotation of the disk in which stars are born had already been recognized by von Weizsäcker some forty years ago, and by the late 1940s a group of young physicists at Göttingen had begun to work on this problem. One of these was Reimar Lüst, now president of the Max Planck Society, whose doctoral dissertation dealt with the transport of the angular momentum in rotating gas disks. In 1979 Tscharnuter was able to prove by computer calculations that a central core, and thereby a star, can form out of a disk when the turbulence of matter redistributes the angular momentum in the disk. Unfortunately, little is known about the turbulent rotation of a gas disk, and consequently, we cannot quantitatively predict how the separation of matter and angular momentum occurs.

This lack of knowledge has temporarily stopped us in our tracks. Before they can move on to the next step, astrophysicists must learn more about the mechanisms that govern the transport of angular momentum through matter. Apparently, it is not just the astrophysicists who do not know what to do with the angular momentum in rotating disks. Nature does not seem too sure either.

A Binary System Is Born

The ring referred to earlier kept on bothering a group of scientists working at our institute. What would happen if nature occasionally knew as little as Bodenheimer and Tscharnuter about how to redistribute the angular momentum, and what would then happen if such rings were actually to form? We have no evidence in nature of rings of matter rotating around an empty center; all we see in the universe are stars and irregular gas masses. So what then happens to the ring?

If we follow the process by computer, we run up against another major problem. The ring that up to now has been axisymmetric now loses that property. To simulate the process new, complicated programs requiring large amounts of storage must be developed. Fortunately, in 1977–78 Tscharnuter, Karl-Heinz Winkler, and Harold Yorke happened to be at our institute at the same time. There they met a young Polish astrophysicist, Michal Rózyczka. The four developed a computer program and discovered what can happen to the Bodenheimer-Tscharnuter ring. Their findings are reproduced in figure 13.3. Over a period of 10,000 years the ring increases in density at two antipodal points, and after 50,000 years these lumps grow into two orbiting clouds out of which two stars can form. The computer has shown us the birth of a binary system.

Perhaps this is an indication that nature can go in two directions. In the one, matter retains its angular momentum, and binary stars are born out of ring structures. In the other, matter and angular momentum are separated, giving rise to central stars with little rotation that are orbited by planets of low mass and higher angular momentum. If that is indeed the case we must conclude that all single stars are accompanied by planets.

Are We Alone?

Even if we do not yet fully understand the path from interstellar cloud to planetary system, the angular momentum of the original matter is beyond all doubt responsible for the birth of planets, and thus for our own existence. In that event all single stars must be surrounded by small planets invisible to us because of their vast distance. But if the planets around the sun are not unique, does that mean that ours is the only inhabited one? Perhaps our Milky Way system is populated by beings of a similar or earlier or more advanced stage of development. Are we alone in the galaxy or does it house other civilized beings with whom we can establish contact?

The OZMA Project and the Arecibo Message

In May 1960 American astronomers at the Greenbank Radio Observatory focused a radio telescope on the star Tau Ceti. At a wavelength of 21 centimeters they sought to find out whether Tau Ceti was sending out radio waves that might possibly be taken for signals of sentient beings. They repeated the process with the star Epsilon Eridani. What made them pick these two stars? One of them is 11 light-years away, and the other 12. True, they are comparatively close to us, but others are even closer. What is important about these stars is that they resemble the sun in temperature, luminosity, and mass, as well as in age. If our sun is surrounded by planets, one of which is harboring a technical civilization capable of building radio transmitters, is it not reasonable to expect Tau Ceti and Epsilon Eridani to have planets with similar civilizations?

Let us assume that their planets are inhabited by beings matching our level of technological development. Would we be able to detect their transmissions? We have been broadcasting into the universe for a long time. Soon after 1945 we sent out radar impulses to the moon and heard their echo. Astronauts on the moon and space probes that penetrate deep into our solar system are guided by radio signals sent out from the earth. We have sent radio impulses to Venus via a radar antenna and recorded their echo. Let us suppose that we have managed to plant such an antenna on a planet orbiting another sun. The 26-meter telescope in Greenbank could receive its signal up to a distance of 9 light-years. The 100-meter telescope in Effelsberg, Germany, could even detect that transmitter up to a distance of 30 light-years. Within that distance from the sun are located 350 stars. If a planet of one of them were transmitting with the technological means available to us on Earth, the Effelsberg radio astronomers could have heard the signals.

Tau Ceti and Epsilon Eridani were tracked from Greenbank for three months, but no signals were detected from them, and so the project named OZMA after the Land of Oz was discontinued in favor of other radio-astronomic observations. The undertaking was also known as Project Little Green Men. Evidently, the Little Green Men chose to remain mute.

And why shouldn't they? Do we on Earth feel any responsibility for maintaining an interstellar information exchange? Are we systematically sending messages to other stars? Aside from one brief transmission on November 16, 1974, nothing much along these lines has been undertaken. On that day a 3-minute message was sent by radio telescope from Arecibo, Puerto Rico, a wide-range transmitter capable of an accurate beam. But what was it to be aimed at? It was decided to send a message to a globular cluster in Hercules. In sites where stars are close together a single message can reach the planets of 300,000 stars. The Arecibo message will arrive at its destination in 24,000 years. If at that time a civilization with a sufficiently powerful radio telescope should be aimed in our direction during just the critical three minutes it will receive our signal. The odds on that are anyone's guess. The Arecibo message was of course intended as a symbolic gesture to mark the reactivation of the telescope after a major overhaul. If we want to keep an eye out for other civilizations, we must listen systematically, and they in turn must transmit systematically.

Among the less systematic efforts to inform others about ourselves are the two engraved, gilded aluminum disks carried into space by the two Jupiter probes, Pioneer 11 and 12 (see figure 13.4). After the completion of their Jupiter mission, the probes will leave our solar system for outer space. Like the Arecibo message the disks contain information about our place in the universe and about us. If these cosmic greeting cards should fall into the hands of sentient creatures, they will learn a great deal about us. But one thing they will never find out—how we look from the rear. That will remain a mystery to them for ever.

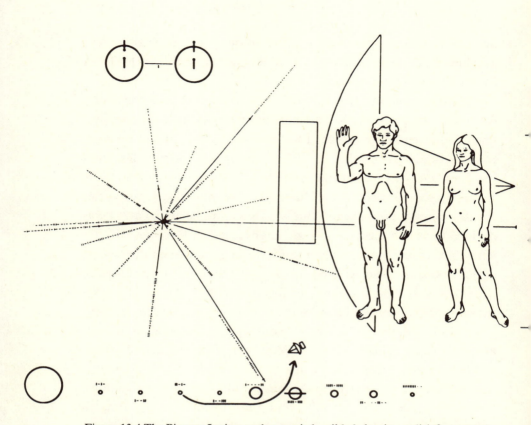

Figure 13.4 The Pioneer Jupiter probes carried a gilded aluminum disk for presentation in possible encounters with extraterrestrial civilizations. In addition to some pictorial information about ourselves the left part of the drawing gives our address in the Milky Way system by giving the directions from which we receive the strongest pulsar signals. The pulsar cycles are in numerals of the binary system. Since the pulsars slow down in the course of time, the receivers can establish the date the probe was sent out. Below, more information about the sun and the solar system, again in numerals of the binary system. (Courtesy of NASA)

The Long Road to Life

Even before we knew fixed stars to be suns, we wondered whether we were alone in the cosmos or whether there was life on and near other stars. Both Nikolaus of Kues (1401–1464) and Giordano Bruno (1548–1600) speculated about the possibility. Kues escaped unscathed, but for this heresy Bruno was burned at the stake.

While investigating the possibility of life on other stars in our galaxy, we will confine ourselves to forms of life whose chemical composition is similar to that of life on our planet. In particular we will assume that it is predicated on the presence of water in liquid form. We will ask whether life similar to ours or even more advanced might have evolved on any other planet, and we will assume that all the planets involved in our inquiry are old enough for life to have evolved on them as it has on ours. We know from the Onverwacht finds in the Transvaal that a relatively highly developed unicellular organism, blue-green algae, already existed 3.5 billion years ago. The earth is estimated to predate these finds by about 1 to 1.5 billion years. A star would therefore have to be at least 4 billion years old to have provided sufficiently stable conditions for the evolution of primitive forms of life.

Let us try to visualize the story of life on our planet. The astronomer Heinrich Siedentopf (1906–1963) worked out a graphic comparison: imagine 5 billion years of the earth's existence compressed into a single year. That would make a week of that year equivalent to 100 million years. A second is thus equivalent to 160 years of actual evolution. Two years have already gone by between the birth of the universe with the oldest star in the Milky Way and the birth of the sun and the earth. Thus, the planets, and with them the earth, have formed in January of the third year. Initially, the atmosphere is still composed largely of hydrogen, the most plentiful element in the cosmos. Only later will it succeed in escaping the gravitational pull of the earth, whose atmosphere will come to be dominated by nitrogen and oxygen. But during the period of hydrogenous atmosphere, simple life has already begun to form, and by March we find the Onver-wacht unicellular organism. Life continues to evolve, but our knowledge of it, based on the fossil record, is limited to the last six months of the compressed year. Hydrogen has largely dissipated, and the living organisms have begun to adjust to oxygen. By the end of November plants, and a little later animals, have taken over the land masses. By Christmas dinosaurs, which had roamed the earth throughout the preceding week, have become extinct. On December 31, at 11 P.M., Peking Man makes his appearance, and ten minutes before midnight Neanderthal Man shows up at the New Year's Eve party; modern man does not make his appearance until five minutes before the new year. Recorded history begins thirty seconds before midnight. Events accelerate most rapidly in the final seconds of the old year; only in the last second before midnight does the earth's population triple. And at about four-tenths of a second before the stroke of midnight, the first radio program is broadcast.

The earth has supported life for the major part of its existence, but only a minute fraction of that life is of a kind we generally refer to as civilized.

Does Our Galaxy Contain a Million Planets on Which Life Exists?

The evolution of life is such a slow process that it can rightly be compared to the evolution of stars. We know that some of the stars in the sky are recent enough for Java Man to have witnessed their birth. Higher forms of life cannot yet possibly have evolved on their planets. We also know that massive stars have been emitting light and heat for only a few million years—far too short a period for life to evolve. Thus, only stars of about 1 solar mass or less are suitable candidates. The Milky Way contains about 100 billion stars, almost all of which are of the right mass. The number of more massive ones is very small.

Except for a small percentage all stars of the Milky Way emit heat long enough to allow for the evolution of sentient life. Whether planets revolve around all the stars remains an open question. Only a body orbiting about a star can have a temperature at which water is liquid. Alas, astronomers know nothing about the solar systems of other suns. Even the stars closest to us are too remote for telescopes to detect possible tiny satellites. Still, it is probable that planets are revolving around other suns; we should not assume that ours is a unique solar system. Over and over again science has proved our belief that our place in the universe is unique to be fallacious.

As we saw earlier the angular momentum of cosmic matter has probably endowed single stars with solar systems. Our own solar system supports this contention. The moons of Jupiter and Saturn constitute mini-solar systems in their own right, most likely also as a result of the angular momentum of the matter out of which they were formed. It is thus reasonable to assume that all single stars are orbited by planets.

But if in the course of stellar evolution, the angular momentum produced a binary system then—if any planets had formed at all—they would after a cosmically speaking short time either have fallen on one of the two main bodies or been hurled out into space. Since at closer inspection more than half of all stars turn out to be binary systems, that leaves only about 40 billion planet carriers.

The question now arises whether the planets are at the appropriate distance from these stars. At least one planet must revolve at a distance such that the radiation reaching its surface gives rise to a temperature at which water is a liquid. As far as our own solar system is concerned, Mercury is too close to the sun, and the planets beyond Mars do not receive sufficient heat from the sun. And as to the planets of other stars, they are invisible to us. How can we know which and how many are at the proper distance? Here we resort to an analogy with our own solar system. Our Earth clearly lies inside the life zone, and Mars and Venus are at the borderline. The pictures taken by the Mariner probes show a Mars surface reminiscent of the moon in its eeriness. Although the Martian atmosphere contains water, the

Viking probes that landed on its surface found no trace of living cells. Soviet probes on Venus have measured temperatures in excess of 450° C. This means that Venus cannot harbor life either. It seems as though we are the only inhabitants of our solar system.

Considering all the conditions that must be satisfied to make life on a planet possible, it becomes clear that it is a rare stroke of fortune for a celestial body to have a climate that can sustain life. Michael H. Hart, a scientist at NASA, demonstrated in 1977 that life on Earth would be impossible if we were even 5 percent closer to the sun than we are, and that Earth would freeze over if we were as little as 1 percent farther away. That does not leave our planet a great deal of latitude. Consequently, Hart concluded that our galaxy contains at most a million planets whose external conditions would make possible the evolution of higher stages of life.

Even if the climate of a planet is favorable for a long enough period, does life then actually form there? That is a question for biologists, not astronomers, although astronomers can help find the answer. They know that, with few exceptions, chemical elements are distributed uniformly throughout the universe. The most remote stars of our Milky Way, and even the stars of other galaxies, are composed of the same mix of chemical elements as the sun. There are no phosphorus stars, nor are there mercury clouds. Hydrogen almost invariably is the major component of cosmic matter, followed by helium, and then the other chemical elements. We can assure the biologists that even on a remote planet with favorable climate they will find the materials needed to build all their organic molecules. Radio astronomers have discovered a great number of different organic molecules in gas clouds—among them alcohol and formic acid, prussic acid and dimethyl ether. Of course, there is still a considerable gap between these simple organic combinations and the more complex molecules that form the basis of that which we call life. Still, suppose that life *does* in fact form wherever it theoretically *can* form; if so, there are a million planets in our galaxy that contain life lasting perhaps as much as 4 billion years. Of course the life on each of these would be at a different stage of evolution.

How Long Does a Civilization Survive?

Inhabited planets are of interest to us only if we can make some sort of contact with them, and here radio signals seem to offer the only possibility. It is therefore only logical for us to ask how many of these one million planets in our galaxy harbor the technical civilization able to send out radio signals. If these planets have been broadcasting continuously throughout the time that life has existed on them, then there should be almost a million broadcasting planets. But green-blue algae do not send out radio signals, nor do forms of life that have destroyed their technology and perhaps even themselves. That leaves only a small number. Our one million planets are thus reduced in number in proportion to the time in which a civilization can send out radio signals measured against the time that life exists on them.

That brings us to the point of greatest uncertainty. The only thing we have to go by is our experience with our own civilization. The technological means of broadcasting into space are of recent origin. But almost simultaneously we have for the first time developed mass means of destruction capable of wiping out all life on Earth. Will we use them? Does a technological civilization send out signals into space only for a paltry few years before it goes about destroying itself?

And *we* ourselves have not even yet begun to broadcast into space. No scientific programs have been organized to send out regular, targeted signals. But let us be optimists and assume that a civilization can solve its problems. Let us assume it lives in peace and prosperity for a million years and hence can afford the luxury of and develop an interest in sending out powerful radio signals into space throughout all that time. That would mean that of the million stars that sustain life only a fraction

$$\frac{1 \text{ million years}}{4 \text{ billion years}}$$

would be transmitting. That would mean that among the one million planets of our galaxy carrying life 250 planets would be sending out signals now. Let us suppose that these planets are evenly distributed throughout our galaxy, making the median distance between two transmitting civilizations around 4,600 light-years. Our signal would travel for 4,600 light-years before it arrived at the nearest transmitting civilization, and it would take another 4,600 years for the answer to come back. Thus, it obviously makes no sense to take a bearing on two nearby stars like Tau Ceti and Epsilon Eridani. It is highly improbable that they have transmitting planets. The only thing that would make sense is to listen for signals from *all* solarlike single stars up to 4,600 light-years away.

It is fewer than four thousand years since the building of the Tower of Babel. But if a civilization lives and transmits only for that brief a period, then the application of the above formula shows that in our galaxy of the million planets with life only the fraction

$$\frac{4{,}000 \text{ years}}{4 \text{ billion years}}$$

is transmitting today. The result: *one* planet. That means that at present except for us at most one other civilization in the entire galaxy would be capable of sending out signals. But if a civilization broadcasts for only a thousand years or less, our efforts to listen in on our galaxy with our radio telescopes are in vain.

Our estimate of the number of planets of our galaxy that may be sending out radio signals rests on many imponderables. I have not tried to calculate their number precisely but rather to demonstrate the factors involved. Our thought exercise has shown us that the greatest uncertainty is related to the fact that we do not know how long a technical civilization can survive. How long does a civilization continue after it has successfully broadcast its first radio wave? A century? Can it survive despite or because of its technical skills?

In posing the question about life in other parts of our Milky Way we have come back to the question of how we here on Earth can survive.

APPENDIX A

The Velocities of Stars

Were it not for spectral analysis our knowledge about the universe would be far scantier. Without it we would know nothing about the chemical composition of stars and only very little about their motion. How does the stellar spectrum help us determine the velocity of a star's movement along the line of sight, that is, its movement toward or away from us? The component of a star's velocity toward or away from the observer is called *radial velocity,* and the phenomenon that makes it possible for us to determine it is the so-called *Doppler Effect,* named for the Austrian physicist Christian Doppler (1803–1853).

Starlight refracted through a glass prism separates, and that separation differs in magnitude for light of different frequencies; blue light, which has a high frequency, is more strongly refracted than red light, which has a lower frequency. If one sets the prism in front of a camera, the resultant picture—the *spectrum* of the star—will show an elongated track instead of a dot. The density of portions of that track is caused by light of different frequencies. The spectrographic instruments of modern astronomy utilize this principle. To investigate faint stars the stellar light is collected by powerful telescopes before it reaches the prism. Instead of prisms other devices are often used to refract light of different frequencies (i.e., light of different color) by differing amounts. Although our camera shows the spectrum as a narrow track, the spectrograph widens it to a strip in which the details emerge more clearly (see figure A.1). The significance of stellar spectra lies in the fact that the atoms of the stellar atmosphere "swallow" light of specific frequencies, and that light is then missing from the spectrum: the strip photographed by the spectrograph shows "lines," that is positions corresponding to specific frequencies at which the photographic emulsion receives no light. This light has been absorbed by the atoms of the stellar atmosphere; the dark lines are called *absorption lines.* Since every atom produces a characteristic group of absorption lines, the chemical composition of a stellar atmosphere can be determined by its spectrum. This forms the basis of the chemical analysis. Everything we have said in this book about the chemical composition of stellar atmospheres and interstellar gas is based on the measurement of spectral lines. The absence of deuterium in the sun and the rareness of lithium were established by just this method. Since the lines of each type of atom appear at different intensities at different temperatures the spectra are also good indicators of the surface temperatures of the stars mentioned in chapter 2. The intensities of the different spectral lines depend sensitively on the temperature prevailing in the chemical composition of a given stellar atmosphere. Thus, stellar spectra are reliable indicators of the surface temperatures of the stars as used in chapter 2. I will not go into this any further here, for all we are interested in at this point is the *Doppler Effect.*

Light is an electromagnetic wave. The electric field strength at any given point in space fluctuates rhythmically when a light ray passes by, and the field strength maxima and minima fly through space with the speed of light. When a source sends out light of a specific frequen-

Figure A.1 The spectrum of 41 Cygni, photographed by Waltraut Seitter of the University of Münster. The waves bent toward the left or violet end of the spectrum have a higher frequency than those bent toward the right or red end. The dark lines are absorption lines of the various atoms. The two adjacent, almost equally strong lines left of center (designated H and K) come from the calcium atom. They can also be found in figure A.2. (Courtesy of W. Seitter, Astronomical Institute, University of Münster, German Federal Republic)

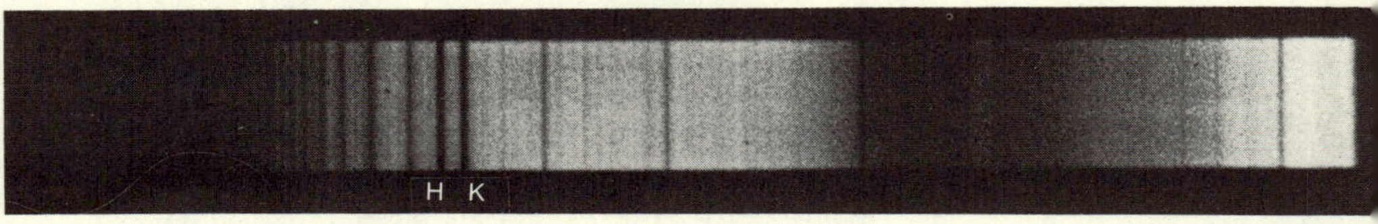

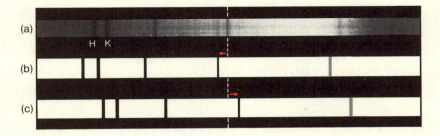

Figure A.2 The Doppler Effect in the spectrum. (a) The original spectrum of a star. (b) Schematic representation of the line shift when the star is moving toward us. All lines have now shifted toward the left, toward the violet range or higher frequencies, shown here by the broken vertical lines and red arrows against the black background. (c) The shift of a star moving away from us. The lines have now moved toward the red end.

cy, we receive it at that frequency only if the distance between source and receiver remains constant. But if the source is moving toward us, the distance traversed by each successive wave maximum is shorter than the preceding one. The wave maxima come to us in more rapid sequence than they were sent out. The light of the source moving toward us appears to be of a higher frequency, that is, bluer, than the light of the same source in the laboratory. Conversely, the light of a receding source appears to be of lower frequency and thus redder than the light of a similar source under laboratory conditions. Actually, it is nothing else but the effect depicted in figure 10.5, for X-ray pulses also vary their periods when the source in its orbit around a star comes toward us and then moves away again.

The Doppler Effect can be measured particularly well by means of the absorption lines of the stellar spectra (see figure A.2). We compare the stellar spectrum with the spectrum of matter, made luminous in a laboratory, whose light is sent through the *identical* spectrograph to see whether the stellar absorption lines of the individual atomic substances in the stellar spectrum are where they are supposed to be or whether they have shifted. With that the determination of the radial velocity of the star thus becomes simple.

The measurement of radial velocity is particularly important in close binaries. During its orbit a star revolving about another star—unless we are looking down on it vertically—moves periodically toward and away from us. This periodic change in its radial velocity can be measured in the spectrum and, as shown in appendix C, can be used to determine the star's mass. In the case of many binary systems, it was established that they were indeed binaries rather than single stars through the line shifts in their spectrum brought about by the Doppler Effect. Because these stars are so far out in space and so close together, the telescope does not resolve them as binaries. Even for noneclipsing binaries, the periodic absorption line shift indicates that we are dealing with two stars orbiting each other.

APPENDIX B

How the Universe Is Measured

We could predict very little about the stars if we did not know the distance between them and us. An insignificant point of light in the sky might be a nearby object less than a meter in diameter producing no light of its own but merely reflecting the light of the sun, or it might be a large body emitting as much light as an entire galaxy, but so far away in the bowels of the universe that the splendor of its radiation is lost to us. It is difficult to draw any inferences about distances in space on the basis of directly measurable distances on Earth.

Measuring our solar system in the age of electronics poses no particular problem, however. All we have to do is take a radar sighting of Venus and apply a law developed by Kepler in the early seventeenth century—Kepler's Third Law—in which he established the connection between the orbital period of the planets around the sun and the radii of their orbits. Let us take two planets, for example Venus and Earth, and designate them *A* and *B,* respectively. According to Kepler we get

$$\text{(orbital period of } A)^2 \times \text{(orbital radius of } B)^3 =$$
$$\text{(orbital period of } B)^2 \times \text{(orbital radius of } A)^3.$$

Since the orbital periods of the planets can be observed directly

(Earth, 365.26 days; Venus, 224.70 days), this formula gives us one equation for the two orbital radii.

It is possible to send a radar signal from Earth to Venus that after being reflected there can be retransmitted to Earth. The travel time of the radar signals moving with the speed of light give us the distance between Earth and Venus, that is, the difference between the orbital radii of the two planets. This in turn gives us another equation for the two unknowns (the orbital radii of Earth and of Venus) which we are able to solve.

The next step takes us from our solar system to the stars. It involves the use of the *parallax method* which had been proposed by Galileo Galilei but was not applied successfully until 1838, when Friedrich Wilhelm Bessel made use of it for 61 Cygni (as mentioned in chapter 4). Since Earth revolves around the sun once in a year, in the course of that year we see the stars in the sky at different angles (see figure B.1 for a schematic view). The length of the line connecting Earth's position on January 1 to its location on July 1 is known. It equals twice the orbital radius of Earth. The two angles between this connecting line and a star can be measured by an astronomer observing that star on these two days. He thus knows two angles and one side of the triangle of figure B.1. And as we learned in elementary geometry, once we know three elements of a triangle (including a length) we can determine all the others. This means that we can figure out the distance between the star and Earth on January 1 and on July 1. In reality the star is so far away that the fine difference between these two distances is insignificant.

We now know the distance between the star and our solar system. This method makes it possible for us to conduct cosmic surveys up to distances of about 300 light-years. In particular the parallax method was used to determine the distances of all stars in the H-R diagram in figure 2.2. As far as stars farther out in space are concerned, the differences between the directions in which they are seen in six-month intervals are so small that they cannot be measured. In their case the method does not work.

We have available another important method for determining dis-

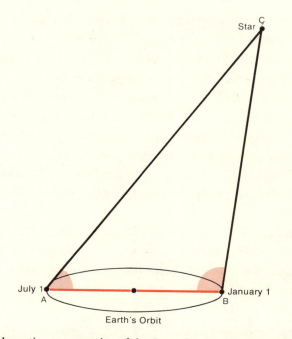

Figure B.1 Schematic representation of the determination of a fixed star by the parallax method. The distance between A and B is twice that of the sun to the earth,
which is determined from the radar echos of Venus. Since the two angles at A and
B can be measured on January 1 and July 1, three elements of the triangle ABC
are known. Calculating the other two sides then becomes a simple matter.

tance that I will touch on only briefly. It is based on the fact that
the velocity of stars in a cluster moving in parallel paths in the same
direction is identical. Even though their movement is barely detectable and measurable, in many star clusters the parallel course of the
stars obviously is aimed toward the same place, like railroad tracks
running parallel toward the same point on the horizon. This point
tells us the direction in which the group of stars is moving. By combining this information with the radial velocity of the stars as measured by the Doppler Effect and with their motion in the sky in front
of more remote stars, it is possible to determine their distance. Again,
these are simple triangular calculations, but I shall not go into any
greater detail here. This was the method used to find the distances
of some star clusters. It was possible to determine their luminosity

and study their patterns in the H-R diagram, as shown in chapter 2.

But the process can be reversed. If a remote star cluster stands so far out in space that the methods described above for determining distance fail, we can make use of the fact that its less massive stars lie along the main sequence and that the luminosity of each corresponds to that of a main-sequence star of its color. Consequently, if I can measure the color of a main-sequence star in a cluster, I know its luminosity. If I then compare this luminosity with its brightness, I can easily find out its distance and with that, the distance of the cluster.

That we are able to penetrate still farther into space borders on the miraculous. For reasons that were not understood for a long time, the pulsating stars of the Delta Cephei type that we discussed in chapter 6 have a peculiar attribute. There is a clear connection

Figure B.2 The period-luminosity relation of Delta Cephei stars. At a specific period, these stars show a specific luminosity. The period is easily determined and thus the mean luminosity as well. This, together with the brightness of the star in the sky, gives us the distance.

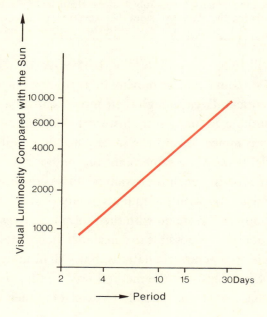

between the cycle of their fluctuations and their mean luminosity (see figure B.2). Since patient observation will yield the pulsation cycle of a Delta Cephei star, the luminosity averaged over a period follows directly from the relationship shown in figure B.2. Because the Delta Cephei stars are highly luminous, their visibility is not only confined to remote crannies of our Milky Way. Their oscillating luminosity makes them stand out even among the stars of other galaxies and through them we have managed to extend our determination of distances beyond the Andromeda galaxy.

APPENDIX C

Weighing the Stars

Even though technology has provided astronomers with refined measuring instruments and modern computers, their methods for determining stellar mass are not much more sophisticated than the three-hundred-year-old techniques of Kepler and Newton. Let me begin with the solar mass. Our planet Earth moves in an almost circular orbit in its gravitational field. It is subject to a centrifugal force that seeks to hurl it out into space. But this force is counteracted by the gravitational pull of the sun, which seeks to draw the earth into the interior of the solar sphere. Earth orbits in just such a way as to balance these two opposing forces. This equilibrium of forces enables us to determine the power of the sun's gravitational pull and thus its mass. We arrive at the answer to it via the formula:

(orbital radius of the planet)^3 = gravitational constant $\times$
(planetary mass + solar mass) $\times$ (orbital period of the planet)^2

The gravitational constant is a number known to us from physics. The orbital radius of the planet Earth can be found by the methods for determining distance described in appendix B. The orbital period of Earth is one year. Consequently, our equation contains only one unknown, the sum of the mass of Earth and sun, and one can solve the above equation for this variable. Since Earth's mass is negligible compared with the mass of the sun, the sum of the two masses is practically the same as the sun's mass alone.

What about the mass of the stars? In the case of binary stars that can be resolved telescopically as two distinct stars revolving about each other, the situation is almost the same, with the sole difference that the masses of the bodies revolving about each other generally are comparable, unlike those of sun and Earth. They spell out much more clearly something we ignored in the above example: the two bodies do not revolve about each other but about a common center of mass. In the case of the two stars of a binary system—let us call them A and B—the following relationship obtains:

$$(\text{distance of the two stars})^3 = \text{gravitational constant} \times (\text{mass of } A + \text{mass of } B) \times (\text{orbital period})^2$$

As to the distance of the two stars from the center of gravity, we apply the following formula:

$$(\text{center of gravity distance of } A) \times (\text{mass of } A) = (\text{center of gravity distance of } B) \times (\text{mass of } B)$$

The distance between A and B naturally is the sum of the distance of both from the center of gravity (see figure C.1). Suppose we were able to see both stars separately in the telescope and could measure their orbits around the common center of gravity. That would give us the distance between them and their orbital period, and we would then know the sum of both masses. We would also see how the stars revolve about each other, which would give us the distance of both from the center of gravity. With that we could use the second equation to give us the ratio of the two masses. Sum and ratio give us the individual masses. As simple as the method appears it still presupposes knowing the distance of the stars from each other, and moreover the orbital radius of each star around the center of gravity. The astronomer may be able to see the orbits of the stars, but he can determine only the angle of their movement in the sky. To find out actual distance he must know the distance of the system from us.

Since we must know the distance of the a binary system in order to determine the masses of the components by the above method, we can apply it only in the case of comparatively close objects. Still it was by this method that the mass-luminosity relationship of main-sequence stars was found (see figure 2.4).

Fortunately we know of another method, and this one does not call for this tedious determination of distance. It is based on the fact that the Doppler Effect (see appendix A) allows us to determine the velocity with which a star moves toward and away from us. If we look at a binary system like that in figure C.1 edgewise, as on the bottom of the figure, and if at a specific time the connecting line between A and B is perpendicular to the line of sight, then the one star is moving toward us and the other is moving away. The velocities

Figure C.1 Motion in a (simple) binary. (top) Looking down vertically on the orbital plane. Both stars, A and B, revolve in different radii around the common gravitational center. (bottom) The same movement seen edgewise. When the connecting line of the two stars is perpendicular to the line of sight—as shown here—then the one star is moving toward us and the other away from us. Their velocities can then be measured with the help of the Doppler Effect described in appendix A.

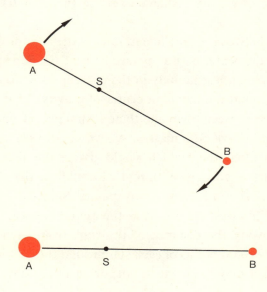

are given by the circumferences of the two orbital paths, divided by the orbital period:

Velocity of $A = \dfrac{2\,\pi \times \text{(center of gravity distance of } A\text{)}}{\text{orbital period}}$

Velocity of $B = \dfrac{2\,\pi \times \text{(center of gravity distance of } B\text{)}}{\text{orbital period}}$

Since the Doppler Effect makes it possible for us to measure both velocities, and since the orbital period can be determined by the rhythm of the motion, the distance of both from the center of gravity follows, and the two previously mentioned formulas give us two equations for the two stellar masses.

The beauty of this method is that the two stars do not have to be seen individually in the telescope. Even when the light of both has fused into a single luminous dot, the spectrum will reveal that the radiation we see comes from two light sources, and the velocities of both of them can be determined separately.

However, in practice matters are more complicated than this. The orbits frequently are not circular but elliptical, and in contrast to figure C.1 we do not look down on them vertically or edgewise but obliquely. But that does not alter the principle of this method.

Since the mass-luminosity relationship of main-sequence stars is known, we have still another method for estimating the masses of many stars. If we know the luminosity of a star and also know it to be on the main sequence, then its mass can be deduced directly from the mass-luminosity relationship. In the case of a main-sequence star of which we know only the surface temperature, we get the luminosity from the H-R diagram, and thus in turn we get the mass from the mass-luminosity relationship. That is helpful when there is no companion to tell us the mass of a star.

Afterword to the
Princeton Science Library Edition

It is now six years since the last American edition of this book appeared. Meanwhile, science has made progress. Some of the open questions of that time have found answers, and in some cases where we thought we already had answers we have learned that we were wrong. The most conspicuous progress, however, has come from new observations. In this afterword I will try to give a summary of the achievements of the last few years in the different fields covered in this book.

The basic ideas about stellar evolution as I had described them are still valid. The rates of the nuclear processes may have changed slightly, but the sun certainly will become a red giant, swallowing the inner planets and coming dangerously close to the Earth. But the case of the missing solar neutrino is still not solved. The chlorine experiment of Raymond Davis in the intervening years gave the same results as his earlier experiment, still finding far fewer neutrinos than predicted by theory. This has also been confirmed by an independent experiment carried out in an old zinc mine in Kamioka, about three hundred kilometers west of Tokyo. There more than two thousand tons of water are observed by photoelectric cells in order to find out whether protons do decay as some modern theories of elementary particles predict. Solar neutrinos also produce light flashes from the B^8 reaction which could be registered and counted. From January 1987 until May 1988 the water was monitored for solar neutrinos with only half as many observed as expected. Meanwhile the search has started for solar neutrinos in the vital reactions for solar energy production. The Gallium experiment GALLEX is under way in a tunnel in the Italian Abruzzi mountain area one thousand two hundred meters deep in the mountain, next to a highway tunnel, exposing thirty tons of gallium to solar neutrinos. Another parallel experiment has been started in the Caucasus. It carries the name SAGE, an abbreviation for Soviet-American Gallium Experiment. Up to now only preliminary results have been published, but it may well turn out that low-energy neutrinos are missing there also. If this is the case then the fault may be in the neutrino and not in the sun. We probably will have to reconsider the properties of neutrinos. Perhaps we have not understood those particles well enough to use them as probes to study the sun.

Astronomers all over the world were excited when on February 23, 1987, a supernova exploded in the Large Magellanic Cloud. Although the event did not occur in our galaxy, it was fairly close, only 170,000 light-years away. It was the closest supernova that has occurred since the invention of the telescope. This time it was possible not only to follow the event in the visible light spectrum with telescopes and spectrographs but also in the ultraviolet and the X-ray region of the spectrum using instruments in orbit. It was also possible to register neutrinos which came from the nuclear processes during the outburst. The experiment in the Kamioka mine, as well as one in a salt mine in Ohio, registered neutrinos that had travelled 170,000 years before reaching the Earth. (Since the Magellanic Cloud is in the Southern Hemisphere, while the detectors in Ohio and in Japan are at northern latitudes, the supernova neutrinos travelled through the Earth's interior and reached the equipment from below.)

Since the Large Magellanic Cloud is a photogenic object, it has been investigated thoroughly in the past. For the first time, therefore, we have pictures of the star before it exploded. The result was a big surprise. On the evening of February 23—the news spread around the astronomical community on the morning of February 24—almost every astronomer would have agreed that all stars which explode as supernovae are red giants. But the star which was present before the explosion of the supernova SN 1987A (as it was named according to the rules astronomers obey when giving names to a supernova) was blue! A contradiction of the basic rules of stellar evolution theory, according to which stars become red giants when they exhaust their hydrogen fuel in their central regions? Not completely, since we have seen that stars can move to the left in the HR diagram, go back to the red region, and then move back and forth several times. It is therefore not completely impossible for a presupernova to be blue, but we do not know more exactly why the star which exploded in the Magellanic Cloud was blue.

Since its discovery the supernova has been observed carefully. Luminous features have been found around the blown-up star. Rings seen around it are due to the sudden outburst of radiation that illuminated the matter in the neighborhood, matter which sometimes has been ejected by the star in earlier epochs and which is overtaken now by the flash from the later explosion. We have observed gamma radiation, which comes from the radioactive element cobalt, precisely from the isotope Co^{56}, which is formed in the exploding material and decays within 114 days. We are eager to know whether a neutron star has formed in the outburst. It would give its secret away if it would send pulsar signals to us, but only if the orientations of its axis of rotation and its magnetic axis would allow the signals to sweep over our planet. But even if the beam of the lighthouse would never send radio waves to the Earth, the region around the pulsar should emit observable X rays in all directions. So far there is no evidence of a rapidly spinning neutron star in the central region of the cloud at the point of the explosion.

During the last years the number of known pulsars went up to more than five hundred. Some were even detected in the Magellanic Clouds. It is surprising how many short-period pulsars are among the new ones. In November 1982 five radio astronomers working with the Arecibo radio telescope found a pulsar which emits 642 pulses per second! This neutron star must rotate more than six hundred times around its axis in each second. Keep in mind that a neutron star contains one to two times the mass of the sun! We might ask why such a fast-spinning object is not torn apart by its centrifugal forces. But if we estimate the gravity at the surface of a neutron star we find that even such a fast rotation produces only a moderate oblateness.

We know that pulsars slow down. In 75,000 years the Crab pulsar will only send fifteen pulses to Earth, not thirty like today. Does this mean that a milli-second pulsar (as a pulsar with hundreds of pulsars per second is called) is extremely young? Yet we have found many millisecond pulsars in globular clusters; most globular clusters are at least ten billion years old. All massive stars which have undergone supernova explosions must have left behind their pulsars billions of years ago, and these neutron stars must have slowed down their rotation. So from where do the rapidly spinning pulsars come? Could there be a mechanism that speeds up old pulsars? If, for instance, a pulsar is a member of a binary system, could the mass flow from the companion to the neutron star, similar to the flow of material indicated in figure 9.8? The arriving matter would increase the rate of rotation. Indeed, we have observed pulsars which show the same effect as discussed in figures 10.4 and 10.5 for X-ray stars. They therefore are members of binary systems. In one case the orientation of the orbit of the neutron star is such that within intervals of nine hours and ten minutes the pulsar regularly disappears behind the companion for forty-four minutes.

Since the last American edition of this book, many efforts have been made to detect radio signals from extraterrestrials. Up to now there has been no success, as everybody knows, since the news would have made headlines. In the process, though, we have learned more about the formation of planetary systems around other stars. The IRAS satellite, and American-Dutch enterprise to investigate infrared radiation from the sky, detected a disk around the

Index

eclipsing variable, 16
Eddington, Arthur, 42–45, 62, 105, 108
Eggen, Olin J., 219
electron, 40
Elsässer, Hans, 207
Emden, Robert, 62
evolutionary track (path), 74, 75, 83, 98, 100

Fabricius, David, 120
falling stars, 10, 117; *see also* meteors
Faulkner, John, 114
Friedman, Herbert, 173

galaxy (galaxies), 3, 4, 224–229, 216–219, plate II; in the Hunting Dogs, 224–228, plate II; *see also* Andromeda galaxy
Galilei, Galileo, 252
gallium neutrino experiment, 90, 91
gamma-ray astronomy, 149
Gamow, George, 45, 47
gas pressure, 57, 105, 106, 209
Giacconi, Riccardo, 172, 173
Giannone, Pietro, 168
giant stars, 18; *see also* red giants
globular star clusters, 30, 31, 33, 83, 115, 191, 192, 218, 219, 239
Gold, Thomas, 144, 145, 150
Goodricke, John, 104, 152
granulation, 60, 61, 65
gravity, 57, 106, 207, 208

halo of the galaxy, 214, 217, 219
Härm, Richard, 113
Hartwig, Ernst, 122, 123, 165
Hayashi, Chushiro, 93
Heckmann, Otto, 123
Heisenberg, Werner, 45
helium flash, 113, 217
helium fusion, 54, 55, 97, 113
Helmholtz, Hermann, 11
Henyey, Louis, 93–95

Henyey method, 93, 113
Hercules X-1, 175–180
Hertzsprung, Ejnar, 22
Hertzsprung-Russell diagram, *see* H-R diagram
Hewish, Anthony, 127, 129, 130, 149, 150
Hipparch, 13
Hirsh, Richard F., 172
Hoffmeister, Cuno, 124, 178, 179, 187
Hofmeister, Emmi, 95
Hohl, Frank, 223
Hopman, Josef, 17
horizontal branch, 33, 103, 114, 115
Houtermans, Fritz, 46, 58
Hoyle, Fred, 80, 93, 130, 144
H-R diagram, 22–34, 36–38, 67, 75, 83, 98, 101–103, 115, 156, 159, 161
Hulse, Russell A., 148
Hyades, 30, 32, 37
HZ Herculis, 178–180

Iben, Icko, 100
Ibn, Butlan, 125*n*
infrared radiation, 213
initial main sequence, 66, 69, 97
initial stars, 67, 211
initial sun, 63, 73
interstellar matter (interstellar gas, interstellar dust), 6, 217, 226, 227
isotope, 41, 49, 51, 53

Jeans, James, 207
Jungk, Robert, 47

Kant, Immanuel, 4, 232
Kepler, Johannes, 124, 251, 256
Kepler's third law, 251, 256
Kienle, Hans, 123
Kohl, Klaus, 160
Kraft, Robert, 168
Kruit, Piet van der, 224
Kues, Nikolaus of, 241

Index

The Princeton Science Library

Hazel Rossotti   **Colour, or Why the World Isn't Grey**

David Ruelle   **Chance and Chaos**

Henry Stommel   **A View of the Sea: A Discussion between a Chief
                      Engineer and an Oceanographer about the
                      Machinery of the Ocean Circulation**

Hermann Weyl   **Symmetry**